From Geometry to Topology

From Geometry to Topology

H. Graham Flegg
MA, DCAe, CEng, FIMA, MIEE, MRAeS, FRMetS

Reader in Mathematics, The Open University

Crane, Russak & Company, Inc.
New York

ISBN 0-8448-0364-2

Library of Congress Catalog Card Number 74-78155

First printed 1974

Copyright © H. G. Flegg

Published in the United States by
Crane, Russak and Company Inc.
347 Madison Avenue
New York, N.Y. 10017

Printed and bound in Great Britain

Author's Preface

This book has grown out of a course of introductory lectures formerly given to intending honours mathematics students at the University College of North Wales, Bangor. The main purpose of the course was to give a largely descriptive and intuitive survey of some of the concepts encountered in a study of topology.

Many university courses in topology plunge immediately into a formalized and entirely abstract presentation of topological concepts. Students often find it difficult in the early stages of such courses to get any real sort of 'feel' for the subject. Even the more simple ideas become difficult to grasp if their simplicity is obscured by the unfamiliar formality of the language in which they are being presented. On the other hand, there is often insufficient time available for a more intuitive and leisurely approach, since it is important that students should come to terms with the real 'meat' of the subject as quickly as possible. This book has been written with the aim of bridging the gap between intuitive and largely geometrical ideas and the formal study of topology.

The first three chapters of the book provide a link from geometry to topology by considering equivalence classes defined by various suitable transformations in real Euclidean space. The general theme here is that, by increasing the number of 'permitted' transformations, the congruence classes of Euclidean geometry are gradually enlarged and the common topological properties remaining within these enlarged classes are thus eventually highlighted. This follows the pattern of Klein's famous *Erlangen Programme* (see Historical Note, page 168). The next nine chapters are devoted to a largely intuitive presentation of certain selected topics in topology designed to stimulate and enlarge the imagination whilst, at the same time, making the fullest possible use of reasonably familiar concepts. Chapter 13 enlarges on the important underlying concept of *continuity* and introduces the concepts of *neighbourhood* and *distance* in an intuitive way so as to be able to discuss the traditional ε–δ approach to the continuity of a function at a point. However, at this point it becomes clear that the minimal use of set theoretic concepts and language adopted so far will no longer be adequate for a more formal presentation of topological ideas, and so necessary set concepts are introduced in Chapter 14 so that the following chapter can discuss functions representing the rigid and elastic transformations encountered earlier in the book. The last two chapters

are devoted to *metric spaces* and *topological spaces* respectively. The aim
here is to formalize the concept of *distance*, which invariably and naturally underlies any initial understanding of transformations of real
Euclidean space, so that the concept of continuity can be freed from
purely geometric overtones and, ultimately, be re-expressed in terms
of *open sets*. The nature of the 'permitted' topological transformations
of earlier chapters can thus eventually be appreciated in their formalized
mathematical context. The final chapter also discusses a few important
properties of topological spaces, and concludes with some particular
remarks about the real number line.

I make no apology for the initial very elementary approach
adopted in this book. The approach has been deliberately elementary
because the book is not intended as a university text-book, but as a
lead-in to a university course for those who would welcome a fairly
informal preview of what the earlier part of a university course in
topology might entail. Indeed, I have had young men and women in
the sixth forms in mind just as much as the first-year university
undergraduate, and I hope also that much of the book will prove suitable
for the more general reader who does not intend taking a formal course
at any time but who is genuinely interested in finding out what topology
is all about. There are no exercises at the ends of the chapters, but a
selection of exercises and problems is provided at the end of the text for
those who want to test their understanding of the material and would
like suggestions for discussion. The chapter on the language of sets has
been specifically included for the benefit of these who are studying (or
have studied) at schools where the 'modern' mathematics syllabuses
have not been (or were not) yet introduced.

Some of the topics introduced are unlikely to be encountered in first
courses in topology, because they belong to the 'fringes' only of the
formal study of topological spaces. These have been included in the
belief that a 'general excursion' is preferable at this stage to a narrow
specialist development. The serious mathematical student should not
be misled, however, into thinking that these sidelines are part of the real
core of the subject, attractive and interesting though some of them may
be. This book is not meant to be a substitute for a serious formalized
study of topological ideas; it is intended to do little more than give
a list of some of the *dramatis personae*, to indicate where the 'plot'
might possibly lead, and to raise the curtain on the scene for the
first act.

H. GRAHAM FLEGG

BANGOR 1969,
MILTON KEYNES 1973

Acknowledgements

I should like to acknowledge my indebtedness to Professor R. Brown, Dr. W. Martyn and Mr. M. Reeves who read the manuscript of this book and made many helpful suggestions, and to my colleague Peter Strain who read the proofs. Such errors and omissions as remain are entirely my responsibility and are in no way attributable to them.

I should like to express my gratitude to the Publishers for their great patience when heavy pressure of work at the Open University meant repeated but unavoidable delays in completing the necessary revision of the original manuscript, and also for producing the book in such a presentable format. My thanks are also due to the illustrators for their care and ingenuity in creating the figures which illustrate the written material, and to Mrs. Barbara Kehoe for transforming sheets of often untidy handwriting into an accurate typescript.

Finally, I must record my thanks due to my friend and former colleague, Charles Barker, from whom I received the original inspiration for the course of lectures at Bangor out of which this book has eventually grown.

H. G. F.

Contents

1

Congruence Classes

What geometry is about—congruence—the rigid transformations:
translation, reflection, rotation—invariant properties—congruence
as an equivalence relation—congruence classes as the concern of
Euclidean geometry.

The traditional study of *geometry* is concerned with certain properties
of figures in Euclidean space. For example, consider the triangle of

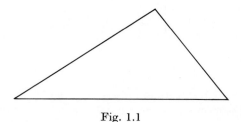

Fig. 1.1

Figure 1.1. This triangle has certain properties such as:

> the values of its angles,
> the lengths of its sides,
> the number of sides,
> its separation of a plane surface into a region inside and a region
>   outside its perimeter,
> the length of its perimeter,
> the area enclosed by its perimeter,
> its orientation with respect to some given axes in space,
> its colour.

Not all these properties are geometric, and, in order to determine
which are and which are not, it is necessary to introduce the concept of
geometric equivalence, often termed *congruence*.

    Intuitively, two plane figures are *congruent* if and only if one may
be placed on top of the other so as to coincide perfectly. The properties
which are shared by every figure congruent to a given figure are

geometric properties. Clearly, all but the last two of the properties listed above are geometric.

The operation of placing one plane figure upon another needs more precise definition. The triangle of Figure 1.2, for example, is congruent

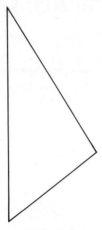

Fig. 1.2

to that of Figure 1.1. Superimposing this second triangle upon the first involves what is known as a *rigid transformation* (or *isometry*). There are three fundamental rigid transformations: *translation*, *rotation* and *reflection*. Every rigid transformation can be expressed in terms of these.

Translation of a point P in a plane is shown in Figure 1.3. If P has co-ordinates (x, y) with respect to the given axes, then the point P' to which it is translated has co-ordinates (x', y') where

$$x' = x+a, \qquad y' = y+b,$$

a being the distance moved in the positive x-direction and b the distance moved in the positive y-direction. (In fact, the figure shows that the transformation of P to P' can be naturally decomposed into two translations, one in the positive x-direction and one in the positive y-direction.)

A plane figure, however, consists not of a single point but of an infinite number of points, though in the case of a triangle three points (the vertices) are sufficient to specify it uniquely. Figure 1.4 shows the translation of a triangle under the same transformation as that of Figure 1.3. Every point belonging to the original triangle is translated by the same amount a in the positive x-direction and by the same

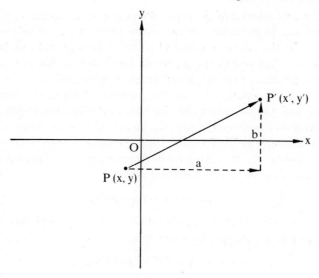

Fig. 1.3

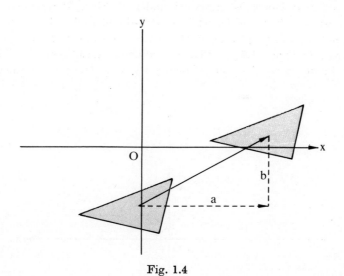

Fig. 1.4

amount b in the positive y-direction. Thus the translation, T say, is given by

$$T : (x, y) \mapsto (x+a, y+b)$$

(which is read as "points (x, y) *map to* points $(x+a, y+b)$"), where the set of all points $\{(x, y)\}$ is the subset of the plane consisting of the perimeter and interior of the original triangle. In a similar way, we can think of any plane figure, or the entire plane itself, being translated under T. In the latter case, x and y would be any real number pair, and the set of all points $\{(x, y)\}$ would be the whole plane, $\mathbf{R} \times \mathbf{R}$ (the Cartesian product of the set of real numbers with itself).

Certain properties, such as the number of sides, the number of vertices, and the separation of the plane into an area inside and an area outside the perimeter of the triangle, are obviously preserved under translations such as T. To show that lengths are preserved, consider any two points P_1, P_2 with co-ordinates (x_1, y_1), (x_2, y_2) respectively. The length of the line P_1P_2 is defined as

$$\sqrt{[(x_2 - x_1)^2 + (y_2 - y_1)^2]}.$$

Under T, the line P_1P_2 is translated to $P_1'P_2'$, say, with co-ordinates $(x_1 + a, y_1 + b)$, $(x_2 + a, y_2 + b)$ respectively. The length of $P_1'P_2'$ is thus

$$\sqrt{[((x_2 + a) - (x_1 + a))^2 + ((y_2 + b) - (y_1 + b))^2]}$$
$$= \sqrt{[(x_2 - x_1)^2 + (y_2 - y_1)^2]},$$

showing that length is preserved under T. Since T represents any translation in the plane, length is preserved under all such translations.

Rotation of a point P about the origin of a plane co-ordinate system is shown in Figure 1.5. If P has co-ordinates (x, y), then P' will have co-ordinates $(x \cos \phi - y \sin \phi, \ x \sin \phi + y \cos \phi)$, where ϕ is the angle

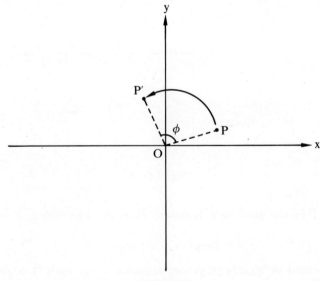

Fig. 1.5

through which the line OP is rotated, as shown, to give OP'. Consider again any two points P_1, P_2 with co-ordinates (x_1, y_1), (x_2, y_2) respectively. The length of the line joining the two points P_1', P_2' to which P_1, P_2 are transformed under a rotation through angle ϕ about the origin is

$$\sqrt{[((x_2 \cos \phi - y_2 \sin \phi) - (x_1 \cos \phi - y_1, \sin \phi))^2 + ((x_2 \sin \phi + y_2 \cos \phi) - (x_1 \sin \phi + y_1 \cos \phi))^2]}$$

$$= \sqrt{[((x_2 - x_1) \cos \phi - (y_2 - y_1) \sin \phi)^2 + ((x_2 - x_1) \sin \phi + (y_2 - y_1) \cos \phi)^2]}$$

$$= \sqrt{[(x_2 - x_1)^2 (\cos^2 \phi + \sin^2 \phi) + (y_2 - y_1)^2 (\sin^2 \phi + \cos^2 \phi) - 2((x_2 - x_1)(y_2 - y_1) \cos \phi \sin \phi - (x_2 - x_1)(y_2 - y_1) \times \sin \phi \cos \phi)]}$$

$$= \sqrt{[(x_2 - x_1)^2 + (y_2 - y_1)^2]},$$

showing that length is again preserved under rotation about the origin. This can be extended to rotations about any point in the plane quite simply. Figure 1.6 shows the rotation of a square in the plane about a point O' with co-ordinates (a, b). This transformation must preserve length, since the axes can be regarded as temporarily translated (as shown) for the purposes of the rotation. O' is now the new origin, and rotations about the origin have already been shown to preserve length. The temporary translation of the axes does not affect the situation, since it has previously been shown that length is preserved under all translations of the plane.

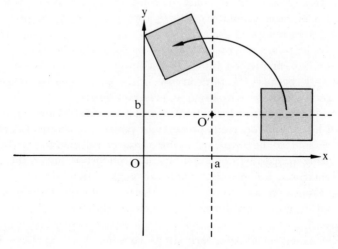

Fig. 1.6

Reflection of two points P_1, P_2 in a given line is shown in Figure 1.7. Rather than repeat a direct formula method for showing that the length of any line P_1P_2 is preserved under reflection, it is simpler first to rotate the whole system about the point of intersection of the given line with the x-axis (or translate the system if the given line and the x-axis are parallel) so that they coincide. The rotation (or translation) preserves

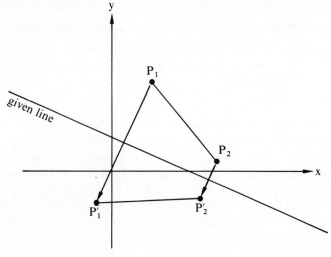

Fig. 1.7

length. It is now necessary only to consider the situation shown in Figure 1.8. If the co-ordinates of P_1, P_2 are (x_1, y_1), (x_2, y_2) respectively, then the co-ordinates of P_1', P_2' are $(x_1, -y_1)$, $(x_2, -y_2)$ respectively; and since, in determining length according to $\sqrt{[(x_2-x_1)^2+(y_2-y_1)^2]}$, the formula is unaffected by the substitution of $-y_1$, $-y_2$ for y_1, y_2 respectively, because the term involving the y's is squared, reflection in the x-axis, and hence in any line, preserves length.

The three rigid transformations, translation, rotation and reflection, thus all have this important property of preserving length. Length is therefore said to be *invariant* under these transformations. Clearly, many other properties of figures are also preserved under the rigid transformations, for example, values of angles, area, the number of sides of a polygon, and so on. One of the most obvious properties not preserved is orientation. Properties which are preserved are said to be *geometric*.

The examples of transformations considered so far have been confined to transformations in a plane. It is not difficult, however, to

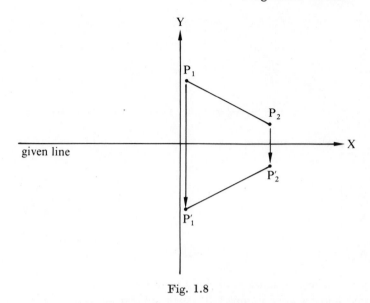

Fig. 1.8

extend the same principles to three dimensions and to consider solid three-dimensional objects. The length of a line P_1P_2 is then defined as

$$\sqrt{[(x_2-x_1)^2+(y_2-y_1)^2+(z_2-z_1)^2]},$$

where the rectangular Cartesian co-ordinates of P_1, P_2 are (x_1, y_1, z_1), (x_2, y_2, z_2) respectively. Indeed, there is no mathematical reason for stopping at three dimensions, and the formula for length clearly has its general counterpart in n-dimensional space. The same extension to three- and higher dimensional space applies to the consideration of invariance under the rigid transformations, though it becomes extremely difficult to visualise what is happening in any space of dimension greater than three.

A space consisting of all points $(x_1, x_2, \ldots, x_n)$ where the distance between $x = (x_1, x_2, \ldots, x_n)$ and $y = (y_1, y_2, \ldots, y_n)$ is defined by

$$d(x, y) = [\sum_{i=1}^{n} |x_i-y_i|^2]^{\frac{1}{2}}$$

is termed a *n-dimensional Euclidean space*. The set of all figures in any n-dimensional Euclidean space can be divided up into distinct subsets such that in any given subset all the figures are equivalent, in the sense that they can be transformed into each other under one or more of the three rigid transformations. Thus the triangles of Figures 1.1 and 1.2 would each belong to one subset, the two squares of Figure 1.6 would each belong to another subset, and so on. The two triangles shown in

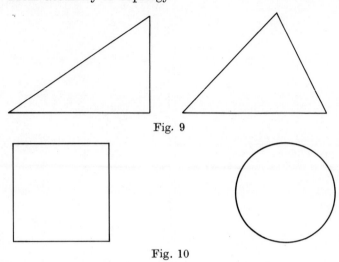

Fig. 9

Fig. 10

Figure 1.9 would, however, belong to different subsets, as would the square and circle of Figure 1.10. Such subsets are termed *equivalence classes*, and the relation

"is congruent to"

on the set of all figures in Euclidean space, which holds for all members of any one equivalence class, is an *equivalence relation*. (An equivalence relation, i.e. a relation having the reflexive, symmetric, and transitive properties, separates the set on which it is defined into disjoint equivalence classes in a unique manner. Thus, if the equivalence relation is changed, the equivalence classes are also necessarily changed.)

In the study of Euclidean geometry, no distinction is made between the members within any one equivalence class. They all share identically the same geometric properties, each is congruent to the other, and hence the equivalence classes of Euclidean geometry are often termed *congruence classes*. To determine that two figures belong to different congruence classes, it is sufficient to find one geometric property which they do not have in common. For example, the triangles of Figure 1.9 do have the same area, since the lengths of their bases are the same and they have the same perpendicular heights. However, they have different angles, and this on its own is sufficient to determine that they belong to different classes, notwithstanding the fact that there are a number of geometric properties which they do share.

Euclidean geometry is thus concerned with the study of classes of figures, and in this context the properties of interest are those which enable it to be determined that two figures belong to different congruence classes by virtue of not sharing any one of these properties.

2

Non-Euclidean Geometries

Orientation as a property—orientation geometry divides congruence classes—magnification (and contraction) combine congruence classes —invariants of similarity geometry—affine and projective transformations and invariants—continuing process of combining equivalence classes.

The individual congruence classes discussed in Chapter 1 can be further divided by taking account, in some way, of orientation in space. For example, in the plane, it may be required that the sides of equivalent polygonal figures make the same angles with some given line. In Figure 2.1 the two triangles are congruent, but in addition they are identically

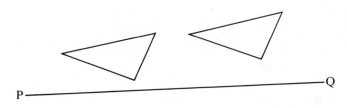

Fig. 2.1

orientated with respect to the line PQ. Triangles not so orientated now belong to different equivalence classes. Within any one equivalence class, the members still share all the same geometric properties, but they share also the non-geometric property of defined orientation. In this new orientated geometry, the only transformation permitted is the rigid transformation of translation.

Free vectors provide an example of a set of one-dimensional 'figures' for which identical orientation is a requirement for equivalence. Thus the study of free vectors involves equivalence classes, within any one of which all the members have the same length and direction (orientation). Members of one such equivalence class are depicted in Figure 2.2. Each individual vector can be thought of as tied to its starting point in space, but, for the purposes of developing a vector

9

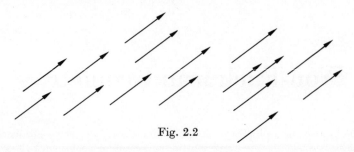

Fig. 2.2

algebra this distinction is ignored, and only the properties common to all, namely length and direction, are considered.

Certain of the geometric congruence classes may, however, be combined by permitting a difference in one or more geometric properties within one equivalence class. For example, it is possible to drop the requirement that lengths should be the same within a class, and to permit transformations which involve proportional magnification (or contraction) in addition to the rigid transformations. In such a geometry, which may be called *similarity geometry*, the two triangles of Figure 2.3 belong to the same equivalence class, and no distinction is made between them.

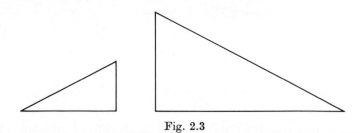

Fig. 2.3

All straight line segments are equivalent in similarity geometry. All squares are equivalent, and all circles are equivalent. Rectangles having the same ratio of side lengths are equivalent, but rectangles of different side-length ratio belong to different equivalence classes. Clearly, area is no longer an invariant under the permitted transformations, but a considerable number of geometric properties are nevertheless retained. In particular, values of angles are preserved, straight line segments remain straight line segments (though their lengths are proportionately changed), and overall 'shape' is preserved without distortion. In three dimensions, no distinction is now made between spheres of differing radii, nor between cubes of differing edge lengths. Certain of the congruence classes of ordinary geometry have now been combined. Congruent figures are indeed still equivalent, but so are all figures which in terms of geometric properties would merely be classed as *similar*.

The pattern which is beginning to emerge is that by increasing the number of permitted transformations, equivalence classes of figures are combined as certain properties cease to be invariant. At each particular stage, it is the study of the invariant properties which forms the basis of the appropriate 'geometry'. This process may now be continued by permitting more and more transformations. For example, in the plane the transformations given by

$$(x, y) \mapsto (ax+by+c, \ dx+ey+f),$$

where a, b, c, d, e, f are real numbers and $ae \neq bd$, preserves neither length, nor angle, nor 'shape'. The geometry which now results is known as *affine geometry*, and its equivalence classes are combinations of equivalence classes of similarity geometry. In affine geometry the two triangles of Figure 2.4 are equivalent, as are also the two triangles of Figure 2.5.

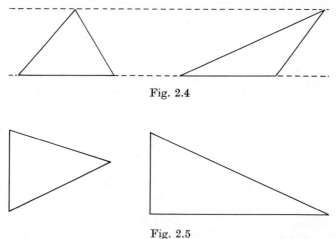

Fig. 2.4

Fig. 2.5

In Figure 2.4, the particular transformation involved, in addition to a translation, is known as a *shear*. The two triangles have the same base length and the same perpendicular height, but the upper vertex has been moved along a line parallel to the translated base. In Figure 2.5, the particular transformation involved, in addition to a translation, is known as a *strain*. Again each triangle has one side which is merely a translation of a corresponding side in the other, but following a translation the remaining vertex has been moved along a line not parallel to the common side. This can be seen more clearly in Figure 2.6, which depicts the strain transformation alone.

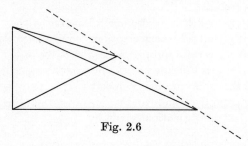

Fig. 2.6

In the case of shear, it can be seen from Figure 2.4 that it so happens that the areas of the two triangles are the same. It is not generally true, however, that area is preserved under affine transformations, as can be immediately seen from Figures 2.5 and 2.6. Indeed, since magnifications and contractions are permitted as in similarity geometry, area cannot be an affine invariant.

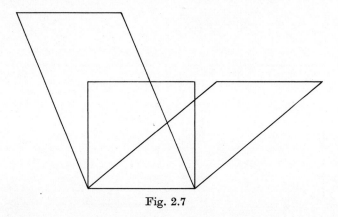

Fig. 2.7

Figure 2.7 depicts a square transformed under shear and also under strain. The two resulting figures are each equivalent to the original square and to each other. Thus, no distinction is made between squares and parallelograms. Further, no distinction is made between circles and ellipses. There are, however, a number of very important properties which are preserved under affine transformations.

Reference to Figure 2.7 shows that under shear and strain, lines which were originally parallel remain parallel although angles between lines are not invariant. It is not difficult to show that this is generally true under all transformations of the plane defined by

$$(x, y) \mapsto (ax+by+c, dx+ey+f),$$

$ae \neq bd$. If PQ and RS are parallel, and P, Q, R, S have co-ordinates (x_1, y_1), (x_2, y_2), (x_3, y_3), (x_4, y_4) respectively, then the equality of their slopes is expressed by

$$\frac{y_2 - y_1}{x_2 - x_1} = \frac{y_4 - y_3}{x_4 - x_3}.$$

Under the transformation, P, Q, R, S map to P', Q', R', S' with co-ordinates $(ax_1+by_1+c, \ dx_1+ey_1+f)$, $(ax_2+by_2+c, \ dx_2+ey_2+f)$, $(ax_3+by_3+c, \ dx_3+ey_3+f)$, $(ax_4+by_4+c, \ dx_4+ey_4+f)$, respectively. The slope of $P'Q'$ is thus given by

$$\frac{d(x_2-x_1)+e(y_2-y_1)}{a(x_2-x_1)+b(y_2-y_1)}$$

$$= \frac{d+e\dfrac{y_2-y_1}{x_2-x_1}}{a+b\dfrac{y_2-y_1}{x_2-x_1}}$$

$$= \frac{d+e\dfrac{y_4-y_3}{x_4-x_3}}{a+b\dfrac{y_4-y_3}{x_4-x_3}}$$

$$= \frac{d(x_4-x_3)+e(y_4-y_3)}{a(x_4-x_3)+b(y_4-y_3)}$$

which is the slope of $R'S'$. This parallel-preserving property of affine transformations means that not all four-sided polygons are equivalent. A square or a parallelogram cannot be transformed into, for example, a trapezium since this would contravene the invariance of parallelism. All triangles are, however, equivalent; no parallel lines are involved, and successive transformations of shear and strain in addition to the rigid transformations will transform any given triangle into any other triangle.

Another important invariant under affine transformations is the ratio in which points divide straight line segments. (A proof of this on lines similar to that for the case of parallelism is not difficult to construct.) A further invariant is that finite configurations remain finite.

These properties are no longer invariants in a geometry, known as *projective geometry*, in which projective transformations are permitted.

Intuitively, such transformations may be thought of as perspective projections of a figure from a point lying outside it. Given two planes in space, not necessarily parallel, then figures in one plane may be transformed into figures in the other either by parallel projection or by projection from an exterior point, as shown in Figures 2.8 and 2.9 respectively. Clearly, the important affine invariant of parallelism is now lost, and hence a square is equivalent to any quadrilateral. However, a straight line remains a straight line under projective trans-

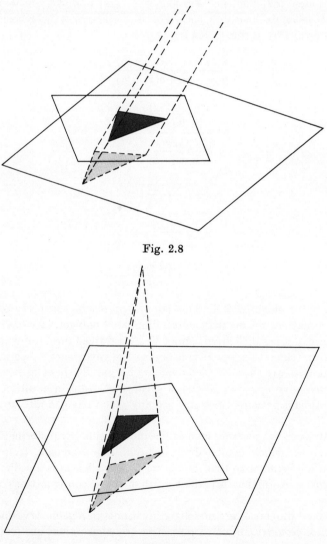

Fig. 2.8

Fig. 2.9

formations, collinearity of points and concurrence of lines is preserved, and finite configurations remain finite.

One particular invariant of considerable importance in projective geometry is *cross-ratio*. If four collinear points P, Q, R, S are transformed under any projective transformation, then, not only are their images P', Q', R', S' collinear, but their respective cross ratios are equal, that is

$$\frac{PR/QR}{PS/QS} = \frac{P'R'/Q'R'}{P'S'/Q'S'}.$$

The case for projection from an exterior point is shown in Figure 2.10. By equating areas of triangles calculated as $\frac{1}{2} \times$ base length $\times$ perpendicular height with those using product of lengths of two sides $\times$ sine of the included angle, it follows that

$$\frac{PR/QR}{PS/QS} = \frac{OP.OR.\sin P\hat{O}R}{OQ.OR.\sin Q\hat{O}R} \cdot \frac{OQ.OS.\sin Q\hat{O}S}{OP.OS.\sin P\hat{O}S}$$

$$= \frac{\sin P\hat{O}R.\sin Q\hat{O}S}{\sin Q\hat{O}R.\sin P\hat{O}S},$$

which remains the same for any four points P', Q', R', S' into which P, $Q.R.S$ may be projected from O. A simple proof, based on similar triangles, can be constructed for the parallel projection case.

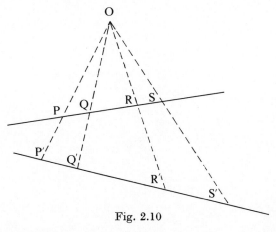

Fig. 2.10

It might seem at first sight that so few of the original geometric invariants now remain that the study of these on their own would not constitute a worth-while 'geometry'. This is far from the case, however. Projective geometry has its own intricate and interesting collection of theorems, which illustrate, amongst other things, the striking principle of *duality* whereby every definition and every theorem holds under the interchange of 'point' and 'line', 'lie on' and 'pass through', 'collinear' and 'concurrent', and so on. In fact, the process of permitting additional transformations and, hence, of combining equivalence classes even further may be valuably continued.

3

From Geometry to Topology

Elastic deformations—intuitive idea of preservation of neighbourhoods—topological equivalence classes—derivation of "topology"—
close connection with study of continuity.

In all the geometries considered so far, one important invariant under
the permitted transformation is the preservation of straight lines as
straight lines. Thus, in none of these geometries does a circle, for
example, belong to the same equivalence class as a polygon. In taking
a further step from projective geometry to *topology*, even this invariant
is abandoned.

In determining which properties of figures are topological, any
one–one bi-continuous transformation is permitted. Intuitively, such
transformations and their inverses map each point to a unique image
point, and points which are 'near' remain 'near', that is, *neighbourhoods* are
preserved. The additional transformations now permitted are sometimes
referred to as *elastic deformations*, and include stretching, bending and
twisting. Cutting is not, however, permitted unless the cut is subsequently 'repaired' in such a way that the 'nearness' of original points is
restored. Joins may not be made in such a way as to bring together
points which were originally separated.

Figure 3.1 depicts some plane figures which all belong to the same
topological equivalence class. Each may be transformed into any of the
others by permitted transformations. If the plane on which the figures
are drawn is thought of as a rubber sheet, it is not difficult to envisage

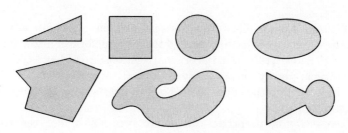

Fig. 3.1

the transformation of, say, the triangle into the square, the square into the circle, and so on.

Figure 3.2 shows a plane closed curve and a knot. These are topologically equivalent, though it is not possible to deform one into the other in three-dimensional space without cutting and subsequent re-joining.

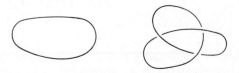

Fig. 3.2

Figure 3.3 illustrates what is meant by the preservation of neighbourhoods under topological transformations. Here, the plane closed curve C is deformed into the plane closed curve C'. P and Q are two points inside C, and R is a point outside C. Under the transformation, P, Q, R are mapped respectively to P', Q', R'. But, even though P', Q' R' are further apart that P, Q, R, the important fact is that P' and Q' are inside C' and R' is outside C'. Indeed, distance apart is irrelevant to the question of 'nearness' in the topological sense, and has no bearing on the preservation of topological invariants under elastic deformations.

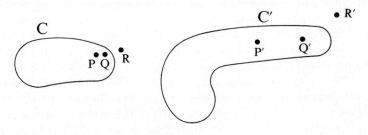

Fig. 3.3

Figure 3.4 again illustrates the preservation of neighbourhoods. Here the circular hole in the centre of the disc is deformed into a narrow slit, but P' and Q' still retain the original relationship of P and Q in that they are separated by the boundary of the hole. A transformation which involves joining up the edges of the hole, that is, 'sewing up' the slit, is not permitted.

The study of properties which are invariant under the transformations now permitted belongs to that branch of mathematics

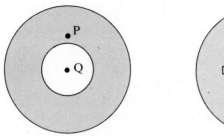
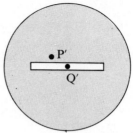

Fig. 3.4

known as *topology*. Because measurement of distance (in the ordinary sense of the word) is not involved, topology may be thought of as the study of *non-metric spatial relationships*. Topological equivalence classes include within one and the same class many figures with widely differing geometric properties. Some properties, invariant under the rigid transformations of ordinary geometry, are still preserved under topological transformations. For example, in the list of properties of a triangle at the beginning of Chapter 1, the property of separating a plane surface into a region inside and a region outside the perimeter of a triangle is included. This property is a topological invariant even though the perimeter of a triangle is now equivalent to any non-self-intersecting closed curve in a plane.

The word 'topology' is derived from the Greek words τοπος meaning *place* and λογια meaning *study*. At one time topology was known as *analysis situs*, a Latin name emphasising its concern with 'situation'. Traditionally, the properties of surfaces in Euclidean space formed a major part of topological study. More recently, however, topology has come to be very largely identified with the study of *continuity*, and it is now regarded as fundamental to a proper understanding of the branch of mathematics known as *analysis* and, in particular, of the limiting processes of *calculus*. It is the fundamentality of topology that is the key to its importance as a corner-stone of modern mathematics. (For a brief account of the genesis of topology as a branch of mathematics see the Historical Note, page 168.)

4

Surfaces

Surface of sphere—properties of regions, paths and curves on a sphere —similar considerations for torus and n-fold torus—separation of surface by curves—genus as a topological property—closed and open surfaces—two-sided and one-sided surfaces—special surfaces: Moebius band and Klein bottle—intuitive idea of orientability—important properties remain under one–one bicontinuous transformations.

Figure 4.1 depicts a *sphere* on whose surface a continuous non-self-intersecting closed curve C has been drawn. It is readily seen that the curve C separates the surface of the sphere into the two distinct regions R_1 and R_2. The regions R_1 and R_2 are said to be *distinct* because it is

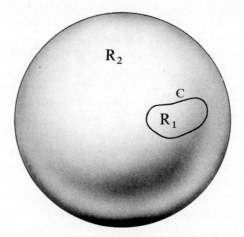

Fig. 4.1

not possible to travel on the surface from a point inside one region to a point inside the other without crossing the curve C. This is true of all points of R_1 and R_2. (For the moment, points lying on C itself, and therefore not strictly in either regions R_1, R_2 are being discounted.)

In Figure 4.2, the points P_1 and P_2 belong to regions R_1 and R_2 respectively, and therefore cannot be joined by any curve on the

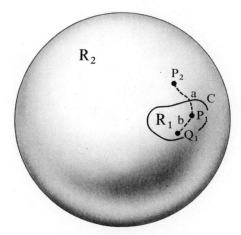

Fig. 4.2

surface not crossing C. The path a illustrates the simplest way of joining P_1 and P_2. This path crosses C once. The points P_1 and Q_1, however, both belong to R_1 and may therefore be joined without crossing C, as illustrated by the path b. Similarly, the points P_2 and Q_2 both belong to R_2 and may also be joined by paths not crossing C. Nevertheless, two points belonging to one and the same region may be joined by paths which do cross C, but in every case it is clear that such paths must cross C *an even number of times*, since any path joining two such points must eventually return to the region in which it started. A similar argument leads to the conclusion that a point belonging to one region may be joined to a point belonging to the other by paths which cross C *an odd number of times*. The path a', for example, in Figure 4.3 crosses C three times. The path b' crosses C twice, and the path b'' crosses C four times. It is, of course, possible to envisage paths which, at some point or other, touch the curve C but do not actually cross it. Provided that such a touching of C is regarded as an even crossing, such paths do not need any special consideration here.

Another property of the continuous non-self-intersecting closed curve C is that it may be gradually contracted on the surface of the sphere so that the smaller of the two regions decreases continuously in size until ultimately the curve shrinks into a point. The important facts here are that the process of contraction may be continuously performed, and that the process may be continued to the limiting stage where the curve C is finally reduced to a single point. This property holds for all continuous non-self-intersecting closed curves on the surface of a sphere.

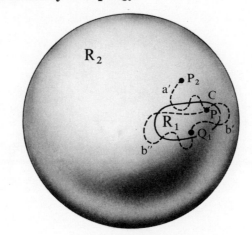

Fig. 4.3

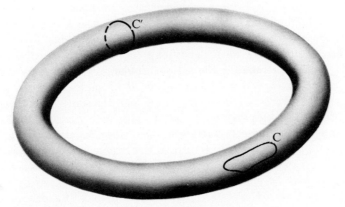

Fig. 4.4

Figure 4.4 depicts the surface of a *one-fold torus*. This has the form of the surface of a traditional ring-doughnut, or of the inner-tube for a car or bicycle tyre. Two distinct non-self-intersecting closed curves C and C' are shown drawn on the surface. Examination of the curve C shows that it has exactly the properties of the curve C previously drawn on the surface of the sphere, namely that it separates the surface on which it is drawn into two distinct regions and that it may be continuously contracted on the surface into a point. The curve C', on the other hand, has different properties. The points P and Q in Figure 4.5, although at first sight possibly appearing to be separated from each other by C', may in fact be joined by paths on the surface of the torus which do not cross C'. For example, the path a joins P

and Q but does not cross C'. Indeed, P and Q may be joined by paths crossing C' any number of times, even or odd. Thus, a' crosses C' once, and a'' crosses C' four times.

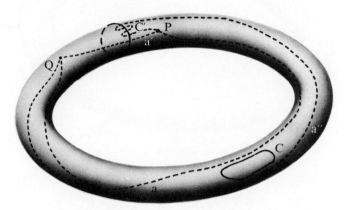

Fig. 4.5

There is, however, another important difference between curves such as C and curves such as C'. No matter how C' is deformed *on the surface of the torus* there is no way of continuously contracting it to a single point. The fact that on the surface of the torus it is possible to draw curves such as C', but not so on the surface of the sphere, provides a means of distinguishing the two surfaces topologically. The sphere and the torus thus belong to different topological equivalence classes. Topologically equivalent surfaces are said to be *homeomorphic*. The sphere and the torus are therefore *not* homeomorphic: there is no one–one bicontinuous transformation by means of which the surface of the sphere may be mapped to the surface of the torus.

If now a further distinct continuous non-self-intersecting closed curve is drawn on the surface of the torus, such as the curve C'' shown in Figure 4.6, then it is easily seen that such a curve inevitably separates the surface into two distinct regions, shown as R_1 and R_2, and it is no longer possible to join points such as P and Q without crossing either C' or C''. Indeed, any path on the surface joining P and Q must make an *odd* total number of crossings of C' and C'': that is, either C' or C'', but not both, must be crossed an odd number of times.

Figure 4.7 depicts the surface of a *two-fold torus*. This may be thought of as a doughnut with two holes. Again it is possible to draw a continuous non-self-intersecting closed curve, such as C, on the surface so that the surface is immediately divided into two distinct regions. The important issue, however, is the number of continuous non-self-

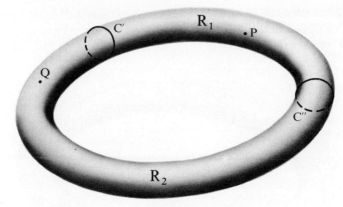

Fig. 4.6

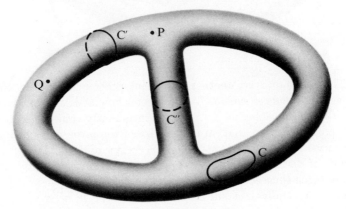

Fig. 4.7

intersecting closed curves which may be drawn on the surface without so dividing it. The curve C' does not, on its own, separate the surface of the two-fold torus into distinct regions. Thus the points P and Q may be joined by paths not crossing C'. If, in addition to C', a curve such as C'' is drawn around the centre 'limb', P and Q may still be joined by paths crossing neither C' nor C''. Similarly, if C'' were to be drawn around the right-hand limb, instead of around the centre limb, the surface would not be divided into distinct regions. If however, as in Figure 4.8, curves C', C'', C''' are all drawn on the surface, then it is inevitably divided into distinct regions and it is not possible to link the points P and Q by any path not crossing at least one of the three curves. It is easily seen that any two of C', C'', C''', taken together, fail to separate the surface of the two-fold torus into distinct regions, but, any two of these having been already drawn, any further continuous non-self-intersecting closed curve drawn on the surface necessarily does so. For the particular points P and Q, shown in Figures 4.7 and 4.8, to be separated so as to belong to different distinct regions, it is, of course, necessary that the third curve be appropriately positioned. The principle, however, is not whether certain prescribed points are specifically separated, but whether or not distinct regions may be identified on the surface. In Figure 4.8, the three curves C', C'', C''' enable the distinct regions R_1 and R_2 to be identified.

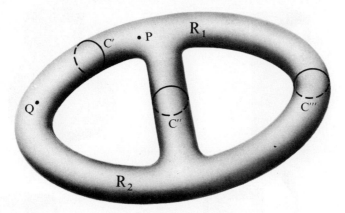

Fig. 4.8

In order to prevent the separating of distinct regions as long as possible, it has been necessary in each case considered to draw continuous non-self-intersecting closed curves which cannot be gradually contracted into points on the surface. In each case, a curve such as C in Figures 4.1, 4.4, and 4.7 immediately separates out distinct regions. In the case of the sphere, it is impossible to draw a curve on the surface which cannot be continuously contracted into a point. The one-fold torus, on the other hand, by virtue of its hole, and also the two-, three-, and n-fold torus for the same reason, all permit the drawing of curves which cannot be contracted.

The greatest number of distinct continuous non-self-intersecting closed curves which may be drawn on a surface without separating it into distinct regions defines the *genus* of the surface. Thus the genus of the surface of a sphere is 0. The genus of the surface of a one-fold torus is 1, of a two-fold torus 2, and, generally, of an n-fold torus n. The genus of a surface is a topological property of that surface; that is, it is invariant under all one–one bicontinuous transformations. All surfaces belonging to one and the same topological equivalence class have the same genus. In fact, for the kinds of surface so far considered (i.e. for closed two-sided surfaces), the converse is also true. Two such surfaces, having the same genus, are topologically equivalent. Thus the genus of such a surface characterizes it completely from a topological standpoint.

Suppose that the torus of Figure 4.4 is now regarded as a solid figure (e.g. as a doughnut). The curve C' may now be taken to define a complete cut made with a knife. The resulting cut doughnut may then be imagined to be straightened out into a solid cylinder, which may

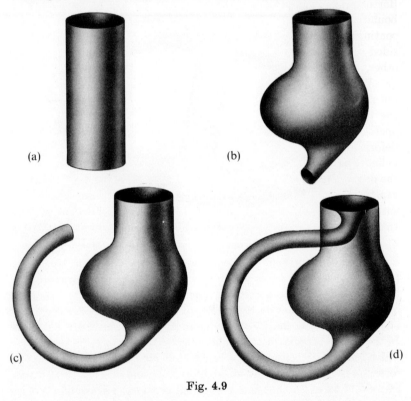

(a)

(b)

(c)

(d)

Fig. 4.9

then be continuously deformed into a solid sphere (or ball). The stages of this deformation are depicted in Figure 4.9 (a) to (f) which is not intended to suggest in any way that a torus may be continuously deformed into a sphere. This is not possible, since the two belong to different topological equivalence classes. Once the solid torus has been cut through, as at C' in Figure 4.9 (a), it then ceases to be a solid torus or to be topologically equivalent to it, and, as the cut is not subsequently 'repaired', the resulting ball cannot be equivalent to the original solid torus. (The cut through at C' also produces in the solid torus two new areas of bounding surface not previously there.)

It is important to appreciate that separations into regions made by drawing curves on a surface, such as that of a sphere or n-fold torus, must not in any way destroy the closed nature of the surface. Intuitively, a surface is *closed* if it has no boundary curves. By this definition, the surfaces of a sphere and any n-fold torus are closed, whilst the surfaces of a hollow cylinder and of a disc are *open*. In this context, boundary curves are not to be taken to include edges of a solid body (for example, the edges of a cube or of a pyramid), nor to include boundaries of regions separated out on a surface by the drawing of continuous non-self-intersecting closed curves. Boundary curves of two-sided surfaces are curves which separate one side of a surface from the other, such as, for example, the edges of a piece of infinitely thin paper.

When the original torus of Figure 4.9 (a) was thought of as a dough-nut rather than as an inner tube, the cut through at C' eventually gave rise to a *solid* cylinder and not to a hollow cylinder, which would be open-ended. The solid cylinder, like the solid sphere into which it is deformed, has a closed surface. The circular edges of the ends of the cylinder are not to be regarded as boundary curves. On the other hand, the drawing of a curve such as C in Figure 4.1, Figure 4.4, or Figure 4.7 can be taken as defining a region which may then be removed from a surface. For this purpose, the torus should be thought of as an inner tube since it is only its surface which is being considered. If, for example, the closed curve C of Figure 4.1 defines a small region which is then removed from the sphere, the resulting surface is open and not closed, and C now becomes a boundary curve locally separating the outside of the new surface from the inside. The new surface obtained will no longer, of course, be topologically equivalent to that of a sphere.

So far, all the surfaces considered have been *two-sided*: intuitively they have the property that a boundary curve must be crossed, where one exists, in order to pass from a point on one side of a given surface to a point on the other side. (If the surface is closed, then it would have to be penetrated in some way.) Figure 4.10, for example, depicts a

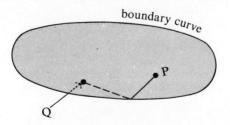

Fig. 4.10

disc with a point P lying on thè upper side and the point Q on the under side. To pass from P to Q it is necessary to cross the boundary curve. Any disc has one and only one boundary curve. A completely open cylinder (open at both ends) has two boundary curves. To pass from a point on the outer side of such a cylinder to a point on the inner side, it is necessary to cross one of the boundary curves an odd number of times. A cylinder which is half-open (open at one end only) has only one boundary curve, and is continuously deformable into, and therefore topologically equivalent to, a disc. Similarly, the removal of a disc from the surface of a sphere leaves an open surface with one boundary curve which is then continuously deformable into, and therefore topologically equivalent to, a disc, and hence also to a half-open cylinder. The removal of two separate discs from the surface of a sphere leaves an open surface with two boundary curves topologically equivalent to a completely open cylinder.

Not all surfaces, however, are two-sided. Figure 4.11 depicts the stages in the formation of a surface by taking a strip of paper and joining up its ends after a 180° twist. The resulting surface is known as a *Moebius band*. Because of the half-twist one side of the original strip

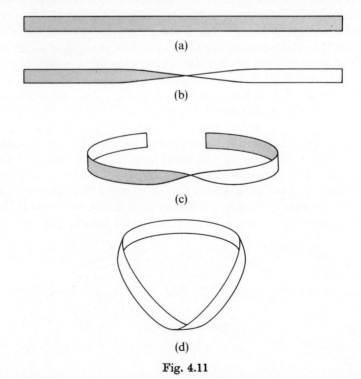

(a)

(b)

(c)

(d)

Fig. 4.11

of paper has been joined to the other side, and the resulting surface is therefore *one-sided*. It is now possible to travel from any point originally on one side of the strip of paper to any point originally on the other side without crossing a boundary curve. In fact, a Moebius band has only one boundary curve, since another effect of the half-twist is that the original opposite long edges of the paper strip have been joined so as to form a continuous curve topologically equivalent to a circle.

There are therefore at least two topologically different types of open surface having one boundary curve only. In all cases, the single boundary curve locally separates points on one side of a surface from points on the other; but, for example, in the case of the Moebius band, looked at overall, there is only one side, so the separation is purely local and does not hold in the context of the band as a whole.

Another example of a one-sided surface is provided by the surface of a *Klein bottle*. Such a 'bottle' is not physically constructable in three-dimensional space. Figure 4.12 conceptually depicts stages in the formation of a Klein bottle. Starting from a completely open cylinder, one end is stretched out, bent over, 'passed through' the curved surface (without breaking or intersecting it), and finally joined up with the other end of the original cylinder from the inside. The operation of 'passing through' the curved surface without breaking or intersecting it cannot be performed in three-dimensional space. It can, however, be 'performed' in an abstract mathematical space of four dimensions. By analogy, the Moebius band, although a two-dimensional surface, cannot be physically constructed in two dimensions only, because of the half-twist. It is not too difficult to see that any two points on the surface of a Klein bottle may be joined by a continuous path not crossing any boundary curve, notwithstanding the fact that from a purely local point of view such points appear to be on opposite sides of the surface. Indeed, the surface of a Klein bottle has no boundary curve whatever. At the final stage of its 'construction', as depicted in the change from (c) to (d) of Figure 4.12, the two open ends of the original cylinder are joined together in such a way that the original outside of the surface is joined to the original inside so that 'outside' and 'inside' may no longer be distinguished. The surface of a Klein bottle is thus both closed and one-sided. (Those familiar with the works of Lewis Carroll will recognise in his *purse of Fortunatus*, as described in 'Sylvie and Bruno', exactly the properties just discussed.)

Closely allied to the property of one-sidedness is the property of *non-orientability*. This is a difficult concept which it is not possible to define fully here. Some intuitive appreciation of what is involved may be obtained, however, from the explanation which follows. If P is any point on a surface and if C is any small continuous closed curve, traced

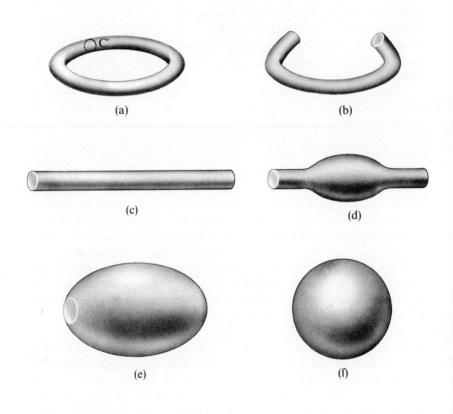

Fig. 4.12

around P on the surface and having a definite orientation, then the surface is said to be *orientable* if the orientation of C is preserved for every continuous closed path traced around P. Otherwise, the surface is said to be *non-orientable*. For example, suppose that P is a given point on a Moebius band. From a purely local point of view there is a corresponding point P' on the other side of the surface. But, since a Moebius band is a one-sided surface, it is possible to draw continuous paths from P to P' without crossing the boundary curve. Such a path is depicted in Figure 4.13. If the small orientated closed curve shown drawn around P is now slid along the path PP', when it eventually arrives at P' its orientation will be reversed.

Another way of seeing that a Moebius band is non-orientable is to consider moving a normal to the surface at P along the continuous path from P to P', so that the foot of the normal is in contact with the path throughout. The normal at P' has the opposite direction to that at P, and yet it has been continuously defined on the surface as it moves along PP'. This is a situation which is not possible with two-sided surfaces.

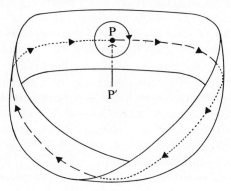

Fig. 4.13

Like the Moebius band, the surface of a Klein bottle is non-orientable. For example, an insect walking about on a Klein bottle may find itself in its original location in space but upside-down relative to its starting orientation (though, of course, this would be a 'mathematical' species of insect capable of taking a four-dimensional walk!). It would appear at first sight, therefore, that the terms 'orientable' and 'two-sided' are synonymous, and the terms 'non-orientable' and 'one-sided' also. For the types of surface considered here there is, in fact, no distinguishable difference between the properties, but at a more advanced stage of study it is sometimes necessary to make a distinction.

A further issue which arises from the discussion of one-sided and two-sided surfaces is that, whereas there is some meaning in distinguishing intuitively between points such as P and P' of Figure 4.13 in the case of two-sided surfaces (because of orientability), there is no reason to make such a distinction in the case of one-sided surfaces. In fact, points such as P and P' are identically located spatially whether the surface be one- or two-sided, so, strictly, P and P' are necessarily one and the same point. The real distinction is not between a point and its opposite point on the other side of a surface, but between a surface which is one-sided and one that is two-sided.

Some of the concepts introduced in this chapter involve taking an unfamiliar view of the generally accepted notion of a surface. The distinction between an open surface and a closed surface is easy to understand and, in the context of the surfaces so far discussed, does not involve any departure from commonplace ideas, even for surfaces such as that of a Klein bottle. Properties such as one-sidedness, however, involve less familiar concepts, but the necessary extension from the commonplace should not prove particularly difficult, and the subsequent discussion on the identification of surfaces with rectangles and other plane figures (to be found in Chapters 11 and 12) should greatly assist in the understanding of the less familiar properties of surfaces.

Distinction has also been made between a *solid*, such as a solid torus or ball, and the surface of such a solid. Care has been taken to specify, for example in the case of a torus or sphere, exactly when it is the surface which is being specifically considered. But by convention, topologists normally take words, such as 'sphere', 'torus', etc. used by themselves, to refer to a surface and not to a solid, and this convention is adopted in the remainder of this book. Whenever a figure is to be understood as a solid, this will be specifically stated.

The really important thing, however, is that there are a great many fundamental and interesting properties left for study even when all one–one bicontinuous transformations are allowed. Figures, including surfaces, may still be classified into their respective equivalence classes by a consideration of their topological properties. Such properties are invariant under the rigid transformations of ordinary geometry, but it is only when the restrictions on permitted transformations are suitably relaxed that their existence is sufficiently highlighted and the study of topology comes into its own.

5

Connectivity

Further topological properties of surfaces—connected and disconnected surfaces—connectivity—contraction of simple closed curves to a point—homotopy classes—relation between homotopy classes and connectivity—cuts reducing surfaces to a disc—rank of open and closed surfaces—rank and connectivity.

A one-sided surface is said to be *connected* if it is possible to travel continuously upon it from any point of the surface to each and every other point of the surface. A two-sided surface is connected if its sides taken separately, are both connected. The sphere and the torus are examples of two-sided connected surfaces. If a disc is separated from a sphere, as depicted in Figure 5.1, then the original total surface will have become *disconnected*. It will, in fact, have been separated into two

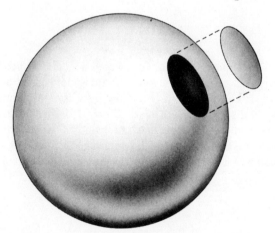

Fig. 5.1

distinct surfaces, each connected in itself and each homeomorphic to a disc. In what follows, the word 'surface' is to be taken as meaning a single connected surface.

It was seen in Chapter 4 that all continuous non-self-intersecting closed curves drawn on a sphere may be contracted into a point. It was also seen that this is a property which does not hold for certain other surfaces such as, for example, a torus. This property may be used intuitively as the basis of the definition of the *connectivity* of a surface,

A surface is said to be *simply connected* if every continuous non-self-intersecting closed curve upon it may be continuously contracted on the surface into a point. Clearly, any surface homeomorphic to a disc or to a sphere is simply connected. Figure 5.2 depicts a continuous non-self-intersecting closed curve C drawn on an *annulus* (the portion of a plane bounded by two concentric circles and topologically equivalent to a cylinder) which cannot be continuously contracted into a point without leaving the surface on which it is drawn. An annulus is thus not simply connected.

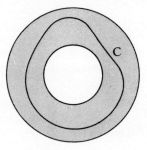

Fig. 5.2

Figure 5.3 depicts an annulus in which a single *cut* has been made. For the purposes of this discussion, a cut is defined as a continuous incision made from a point on a boundary to another point on a boundary which does not disconnect the surface. Clearly, for an annulus, such a cut cannot start and end on one and the same boundary otherwise the surface would be disconnected. The annulus with one cut is topologically equivalent to a disc and is thus simply connected. A surface requir-

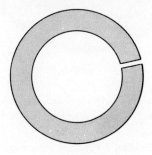

Fig. 5.3

ing only one cut to make it homeomorphic to a disc is said to be *doubly connected*.

Figure 5.4 depicts an annulus with three holes. It is not possible to reduce this to a surface homeomorphic to a disc unless three cuts are made. Such a surface is said to be *quadruply connected*. There are several

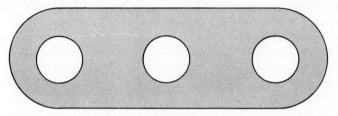

Fig. 5.4

alternative ways of making the three cuts which will render this annulus homeomorphic to a disc. Figure 5.5 depicts the cuts made in each instance from the boundary of one of the holes to the exterior boundary of the annulus. Figure 5.6, on the other hand, depicts the first two cuts made so as to link the holes together. At this point, the surface is now equivalent to the annulus with a single hole, and a final cut from the interior boundary to the exterior boundary will make it simply connected.

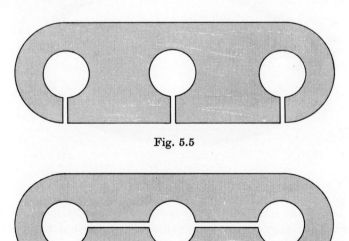

Fig. 5.5

Fig. 5.6

This concept of the connectivity of a surface may now be generalized. A surface requiring $n-1$ cuts in order to render it homeomorphic to a disc is said to be *n-tuply connected*. The cuts are not, of course, allowed to intersect each other. This follows from the definition of 'cut' (given earlier), since, once a cut has been made, it forms parts of a boundary. Crossing an already made cut with another would involve the completion of one new cut and the commencement of another. An *n*-tuply connected surface for which *n* is greater than 1 is said to be *multiply connected*. A further example of such a surface is the curved surface of a cylinder (for which $n = 2$).

In the case of closed surfaces, such as a sphere and a torus, it is not possible to make an initial cut from one boundary to another since these surfaces have no boundaries. However, if it is first imagined intuitively that a 'pinhole' is made in the surface, the process of making cuts may then proceed, so long, of course, as the surface is not thereby disconnected. The two cuts required to make a one-fold torus homeomorphic to a disc are depicted in Figure 5.7. The original surface is triply connected. The first cut made renders it homeomorphic to the curved surface of a cylinder, which is doubly connected. The second cut finally makes it simply connected.

Fig. 5.7

In the case of the sphere, it is not possible to make any cut, following the making of a 'pinhole', which does not disconnect the surface. The sphere thus requires zero cuts, according to the principles discussed above, in order to make it homeomorphic to a disc. However, it is clear from previous discussions that the sphere is not topologically equivalent to a disc, since a disc is open whilst the sphere is closed. The sphere with the 'pinhole' is, however, to be regarded as the starting point, and a sphere with any single hole in it and a disc are homeomorphic.

Because it is possible to contract any continuous non-self-intersecting closed curve to a point on a simply connected surface, it follows that every such curve may be continuously deformed on the surface into any other such curve. This property is expressed by stating that all continuous non-self-intersecting closed curves on a simply connected surface are *homotopic* to each other. Figure 5.8 depicts two closed curves C and C' on a disc. These curves are homotopic to each other since either can first be contracted into some common point, such

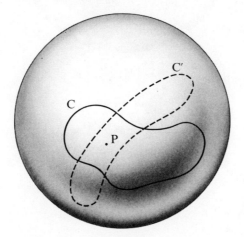

Fig. 5.8

as P, and then re-expanded into the other. Figure 5.9, however, depicts two continuous non-self-intersecting closed curves on a disc which enclose no common point. In this case, it is possible to deform one of the curves so that some common point is eventually enclosed. The conditions of Figure 5.8 then apply.

Figure 5.10 depicts four continuous non-self-intersecting closed curves C_1, C_1', C_2, C_2' on the surface of a torus. In this case C_1 and C_1' are a homotopic pair, and C_2 and C_2' another homotopic pair. These represent two distinct *homotopy classes*, however. C_1 and C_1' are not homotopic to C_2 and C_2'. One distinction between these two homotopy classes is immediately evident. C_1 and C_1' may both be continuously contracted to a point, C_2 and C_2' may not. This does not, however, prevent C_2 and C_2' being deformed into each other. In the case of these two particular curves, such a deformation merely involves 'sliding' one of them around the surface. This may be regarded as a simple trans-

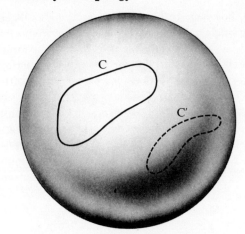

Fig. 5.9

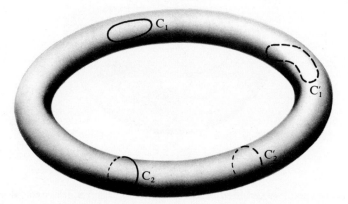

Fig. 5.10

lation. Curves such as C_1 and C_1', which may be continuously contracted into a point on a surface, are said to belong to the *null homotopy class*.

There is clearly a direct relation between homotopy classes on a surface and the connectivity of that surface. A surface is simply connected if every continuous closed curve upon it belongs to the null homotopy class.

The *rank* of an open surface is defined as the least number of cuts required to make the surface homeomorphic to a disc. The rank of a closed surface is defined as the rank of the open surface obtained from the closed surface by the removal of a single disc. The rank of an n-tuply connected surface is therefore $n-1$. Thus, a disc and a sphere have rank zero. A simple annulus (an annulus with one hole), the curved surface of a cylinder, and a Moebius band have rank 1. A one-fold torus and an annulus with two holes have rank 2.

An alternative definition of rank is possible. The rank of a surface may be defined as the greatest number of non-intersecting cuts which can be made in it without disconnection. This alternative definition applies equally to open and closed surfaces.

It is not difficult to see that the two definitions of rank are equivalent. A two-fold torus from which a disc has been removed provides an example demonstrating this equivalence. Such a surface is depicted in Figure 5.11. A first cut having been made, either as shown or in any other way not disconnecting the surface, two further cuts must be made before the surface becomes homeomorphic to a disc. Alternatively, although it is possible to disconnect the surface by a single cut, it is possible to make a maximum of three cuts without disconnecting the surface. If more than three cuts are made the surface is necessarily disconnected. From either stand-point, therefore, the rank of the surface is 3.

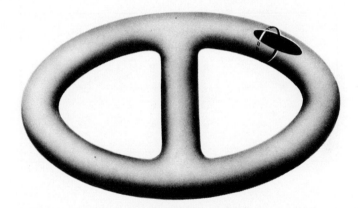

Fig. 5.11

Because of the direct relationship between the *rank* of a surface and its *connectivity*, reference to rank is often omitted in discussion of the properties of surfaces. It will, however, be encountered again in this book during the discussion of the standard model of a surface in Chapter 12.

6

Euler Characteristic

Maps—interrelation between vertices, arcs and regions—Euler characteristic—polyhedra—five Platonic polyhedra—Euler's formula—Euler characteristic as a topological property—relation with genus—flow on a surface—singular points: sinks, sources, vortices, etc.—index of a singular point—singular points and Euler characteristic.

Figure 6.1 depicts five vertices linked together by eight non-intersecting arcs in such a way as to separate the enclosed area into five simply connected regions. Such a separation of a surface is termed a *map*.

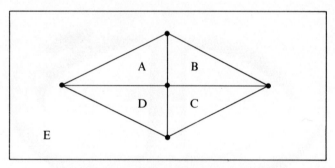

Fig. 6.1

The map may be regarded as drawn on the surface of a sphere, as in Figure 6.2. In both cases, the given map separates the surface on which it is drawn into the five regions A, B, C, D, E. If the number of vertices of such a map is V, the number of arcs (or edges) is E, and the number of regions (or faces) is F, then, for the map of Figures 6.1 and 6.2,

$$V = 5, E = 8, F = 5.$$

This particular map will now be designated M_1, and the corresponding numbers of vertices, arcs, and regions, V_1, E_1, F_1 respectively.

Figure 6.3 depicts a second map, M_2, which has three vertices and three arcs, and which separates the sphere on which it is drawn into two regions. In this case,

$$V_2 = 3, E_2 = 3, F_2 = 2.$$

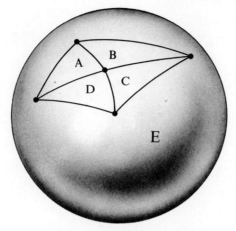

Fig. 6.2

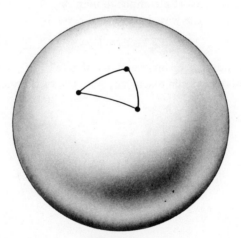

Fig. 6.3

The maps M_1 and M_2 may be superimposed as they are drawn on equivalent surfaces. Such a superimposition gives the composite map M_3 depicted in Figure 6.4. In superimposing two such maps, care must be taken to ensure that vertices of one map have not been placed directly upon vertices of the other, and that arcs of one have not been placed directly upon arcs of the other. For any such composite map constructed from two basic maps, the number of vertices must be equal to the sum of the numbers of vertices of the basic maps together with the number of additional vertices defined by the intersections of the arcs of the two basic maps consequent upon their superimposition. Thus,

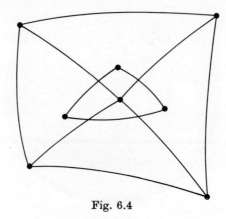

Fig. 6.4

the number of vertices of a composite map M_3, formed from two basic
maps M_1 and M_2 in the way described, is given by

$$V_3 = V_1 + V_2 + v,$$

where v is the number of intersections of arcs of M_1 with arcs of M_2.
In the particular example discussed, $v = 4$, and therefore, $V_3 = 5 + 3 + 4 = 12$.

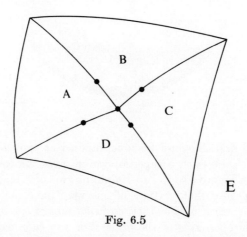

Fig. 6.5

Figure 6.5 depicts the original map M_1 modified by the inclusion
of the vertices defined by the intersections of the edges of M_1 and M_2
when these maps are superimposed on each other. This modified map

may be denoted by M'. Clearly, each additional vertex divides some arc of the basic map M_1 into two separate arcs. Thus, if

$$V' = V_1+v,$$

then

$$E' = E_1+v.$$

On the other hand, the number of regions remains unchanged; that is

$$F' = F_1.$$

For the map M' of Figure 6.5, $V' = 5+4 = 9$, $E' = 8+4 = 12$, $F' = 5$. In particular,

$$V'-E'+F' = (V_1+v)-(E_1+v)+F_1 \tag{1}$$

$$= V_1-E_1+F_1.$$

The composite map M_3 is now obtainable from the modified map M' by adding successive *chains* of arcs, each chain linking two vertices and dividing a region into two parts. If a chain includes n arcs, then it introduces $n-1$ new vertices. At the same time, it increases the number of regions by one. For example, the chain depicted in Figure 6.6 consists of two arcs and introduces one new vertex to the map of Figure 6.5. At the same time it separates the region B into two parts,

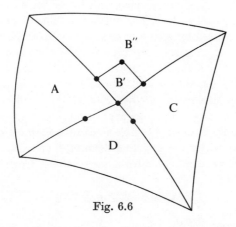

Fig. 6.6

B' and B''. When all the chains required for the completion of the composite map M_3 have been included, the number of vertices V_3 will be given by V', the number of vertices of the modified map, together with the total number of vertices added as a result of the inclusion of the various chains. Similarly, the number of arcs E_3 will be given by E' together with the total number of arcs in all the added chains, and the number of regions F_3 will be given by F' together with the number of added chains. Thus, in particular,

$$V_3 - E_3 + F_3 = V' - E' + F' + \sum [(n-1) - n + 1],$$

where the right-hand term is summed over all the chains added to the modified map. Hence,

$$V_3 - E_3 + F_3 = V' - E' + F'$$
$$= V_1 - E_1 + F_1$$

from (1) above. For the particular maps depicted in Figures 6.2 to 6.5,

$$V_1 - E_1 + F_1 = 5 - 8 + 5 = 2,$$
$$V_2 - E_2 + F_2 = 3 - 3 + 2 = 2,$$
$$V_3 - E_3 + F_3 = 12 - 19 + 9 = 2,$$
$$V' - E' + F' = 9 - 12 + 5 = 2.$$

Since any map on a given surface can be built up by superimposing simple basic maps on each other, it follows that the expression

$$V - E + F$$

is invariant for any given surface, and hence for all surfaces homeomorphic to the given surface. The number obtained from the expression $V - E + F$ is denoted by χ, and is termed the *Euler characteristic* of the surface. It is a topological property of the surface and is independent of any map to which actual values of V, E and F apply.

Clearly, for a sphere, $\chi = 2$. Thus, if any two of the quantities V E, F for a map on a sphere are determined, the third quantity is automatically determined from the expression $V - E + F = 2$. It follows that, for example, it is not possible to draw a map on a sphere having six vertices linked by ten arcs and defining four regions. That χ for a sphere differs from that for a torus may be seen from Figure 6.7. For the map shown,

$$V = 2, E = 4, F = 2;$$

hence, $\chi = 2 - 4 + 2 = 0$.

Any polyhedron, which is both closed and convex, is termed a *simple* polyhedron. Such a polyhedron may be continuously deformed into a sphere, and hence it follows that

$$V - E + F = 2,$$

where V is the total number of its vertices, E is the total number of its edges, and F is the total number of its faces. This must hold since the original vertices, edges and faces simply deform into a map on the surface of the sphere. If, in addition to being closed and convex, a polyhedron has each of its faces congruent to the same regular polygon then it is termed *regular*.

Fig. 6.7

Let a regular polyhedron have each face an n-sided regular polygon, and let f faces meet at each vertex. Since each edge is an edge of two faces and links two vertices, it follows that

$$fV = 2E = nF.$$

However, $V - E + F = 2$; hence, substituting for V and F gives

$$\frac{2E}{f} - E + \frac{2E}{n} = 2.$$

Division by $2E$ and rearrangement gives

$$\frac{1}{f} + \frac{1}{n} = \frac{1}{E} + \frac{1}{2}. \tag{2}$$

Now, the minimum number of faces which meet at any vertex must be three. Similarly, the minimum number of sides of a regular polygon is three. Thus

$$f \geqq 3 \text{ and } n \geqq 3.$$

However, if f and n were both greater than three, the right-hand side of (2) would not exceed one half. Hence, either

$$f = 3 \text{ or } n = 3.$$

When $F = 3$ holds, n cannot be greater than five, since again the right-side of (2) must exceed one half. Similarly, when $n = 3$ holds, the same applies to the value of f. There can therefore be at most five regular polyhedra, namely those for which

$$f = 3, 3 \leqq n \leqq 5 \text{ or } n = 3, 3 \leqq f \leqq 5.$$

Such five regular polyhedra were known in Plato's day, and find a place in the Platonic writings. For this reason, they are frequently termed *Platonic* polyhedra. The five regular polyhedra are listed in the table below, and depicted in Figure 6.8.

f	n	V	E	F	name
3	3	4	6	4	tetrahedron
4	3	6	12	8	octahedron
3	4	8	12	6	cube
5	3	12	30	20	icosahedron
3	5	20	30	12	dodecahedron

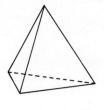

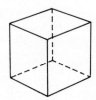

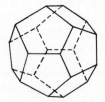

Fig. 6.8

The expression

$$V - E + F = 2,$$

when applied to polyhedra is known as *Euler's formula*. A direct proof of this formula for polyhedra may be obtained using the *method of triangulation*. Starting with a polyhedron having V vertices, E edges, and F faces, it is first necessary to remove one face. The surface so obtained is then deformed until the E edges and the remaining $F-1$ faces lie in a plane and thus may be regarded topologically as a disc. It is important to note that the removal of a face does not involve any reduction in the number of vertices, nor in the number of edges. The process of triangulation is now carried out. Each face is divided by the drawing of diagonals in such a way that at least one triangle is formed each time a diagonal is drawn. The process is continued until only triangular faces remain. Clearly, each time a diagonal is drawn, E and F will both increase by one. Hence, if the total triangulation process requires the drawing of d diagonals, the expression

$$V - E + F.$$

applicable to the original polyhedron, will now become

$$V - (E+d) + (F+d-1). \tag{3}$$

The triangles are now removed one by one. In each instance, the triangle removed must be one which has at least one edge on the original or subsequent boundary. If a triangle so removed has one edge on the boundary, as depicted in Figure 6.9, then after its removal V

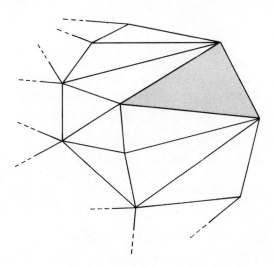

Fig. 6.9

will be unchanged for the remaining figure, but E and F will each have decreased by one. There will thus be no overall change in the value of expression (3). If a removed triangle has two edges on the boundary, as depicted in Figure 6.10, then V and F will each decrease by one, whilst E will decrease by two. Again there will be no overall change in the

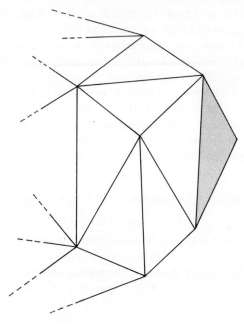

Fig. 6.10

value of expression (3). Ultimately, one triangle only will remain, and the value of expression (3) will still be unchanged. This one triangle will have three vertices, three edges and one face, giving

$$V - E + F = 1.$$

Since the value of $V - E + F$ has not been altered by the triangulation process, it follows that for the original polyhedron with one face removed $V - E + F = 1$ also. Reconstitution of the original polyhedron by the restoration of the missing face increases the value of F, and hence also of $V - E + F$ by one. The validity of Euler's formula for polyhedra is thus established.

An example of the triangulation process is depicted in Figures 6.11 to 6.14. A cube has

$$V = 8, E = 12, F = 6.$$

One of the faces, *EFGH* say, must now be removed and the resulting surface deformed so as to lie in a plane. This is depicted in Figure 6.12. At this stage

$$V = 8, E = 12, F = 5.$$

Triangulation may now be carried out as depicted in Figure 6.13, after which

$$V = 8, E = 17, F = 10.$$

Triangles are now removed from the boundary inwards until only one triangle remains. An intermediate stage is depicted in Figure 6.14. Ultimately only one triangle, *ABC* say, remains for which

$$V = 3, E = 3, F = 1.$$

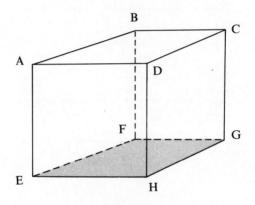

Fig. 6.11

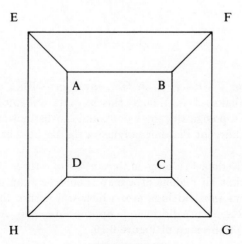

Fig. 6.12

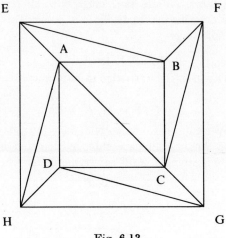

Fig. 6.13

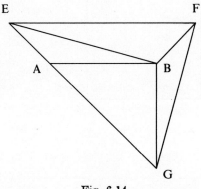

Fig. 6.14

The expression $V - E + F$ is, in fact, invariant within any one topological equivalence class, though this is quite difficult to prove. All
polyhedra are topologically equivalent, and the distinction between, for
example, the different Platonic polyhedra rightly lies outside the study
of topology.

Figure 6.15 depicts a map on the surface of a torus. Because of the
requirement that all regions of a map should be simply connected, a
map on a torus must include arcs which will ensure that this holds.
(The map of Figure 6.7 did include appropriate arcs). Arcs C_1 and C_2
ensure this. For the map of Figure 6.15,

$$V = 4, E = 7, F = 3.$$

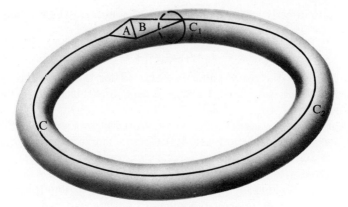

Fig. 6.15

The Euler characteristic is therefore given by $4-7+3 = 0$, as in the case of the map of Figure 6.7.

Figure 6.16 depicts a map drawn on the surface of a two-fold torus. Again, arcs have been included so as to ensure that all regions are

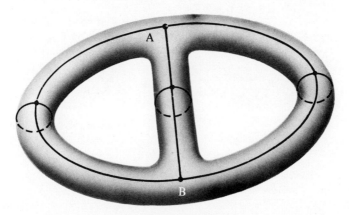

Fig. 6.16

simply connected. For this map,

$$V = 5, E = 9, F = 2,$$

and the Euler characteristic is therefore $5-9+2 = -2$.

Now a sphere has genus $g = 0$, a torus has genus $g = 1$, and a two-fold torus has genus $g = 2$. The number of special arcs which have to be included in any map on a surface which is not simply connected is directly related to the genus of the surface. Hence, the Euler characteristic is also directly related to the genus of a surface. The relationship is given by the expression

$$\chi = V - E + F = 2 - 2g,$$

which confirms the results already obtained for the particular maps which have been considered. Further discussion of this is deferred until Chapter 12.

The Euler characteristic of a smooth closed surface may be obtained from considerations of an entirely different character which rightly belong to a part of the study of topology known as *differential topology*. Suppose that with each point on a sphere there is associated a direction. One way of thinking of this from a purely intuitive standpoint is to imagine that the exterior of the sphere is entirely covered with hair. When the hair is brushed down, so that each hair can be thought of as

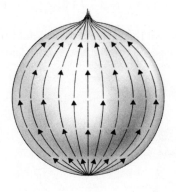

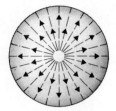

underside showing "hole"

Fig. 6.17

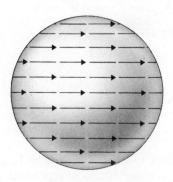

underside showing "crown"

Fig. 6.18

lying on the surface, the direction in which any single hair is lying defines the direction associated with the point at its base. Continuity of direction may be obtained locally by appropriate brushing so that there is no sudden reversal of direction.

It is not possible to brush a 'hairy' sphere in such a way that there is no discontinuity of direction anywhere on its surface. Figure 6.17 depicts a sphere brushed upwards. Continuity of direction on the surface is achieved everywhere except at the top and the bottom of the surface, where there must be a 'tuft' and a 'hole' respectively. Figure 6.18 depicts the sphere brushed horizontally. Again, continuity of direction is achieved everywhere on the surface except at the top and the bottom where there will be 'crowns'. With a 'hairy' torus, however, it is possible to brush the hairs in such a way that continuity of direction is achieved everywhere on the surface. Figures 6.19 and 6.20

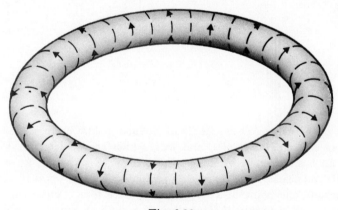

Fig. 6.19

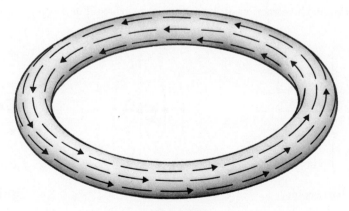

Fig. 6.20

depict two ways of doing this. The fact that it is possible to brush a 'hairy' torus so as to achieve continuity of direction on the whole of the surface whilst it is not possible in the case of the sphere is further evidence of the topological distinction between the sphere and the torus.

An alternative way of considering continuity of direction on a smooth closed surface is to replace the concept of brushing hair on the surface by that of fluid flow. The 'tuft' now becomes a *sink*, the 'hole' becomes a *source*, and the 'crown' becomes a *vortex*. These are depicted in Figure 6.21. Sinks, sources and vortices are examples of *singular*

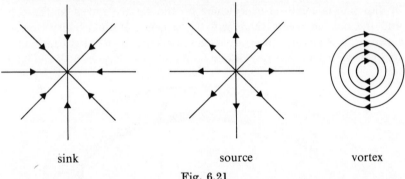

sink source vortex

Fig. 6.21

points on a surface. Other kinds of singular points are also possible, such as *crosspoints*, *dipoles*, and *foci*, which are depicted in Figure 6.22. Each singular point has an integer, called its *index*, associated with it, and this is obtained for any particular singular point by travelling around the point in a circular path once in a counter-clockwise direction and counting the number of counter-clockwise revolutions made by a little arrow with its base on the path and its head always pointing in the direction of the flow on the surface. For a sink, a source,

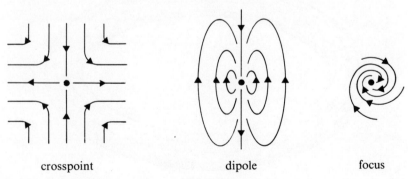

crosspoint dipole focus

Fig. 6.22

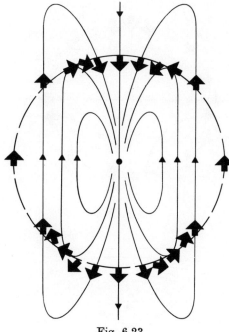

Fig. 6.23

a vortex, and a focus the index is 1. For a dipole, the index is 2, and for the crosspoint depicted in Figure 6.22 the index is -1. Figure 6.23 shows how the index-value 2 is obtained in the case of a dipole. The negative value for the crosspoint arises because one revolution is made by a little arrow *clockwise* as a path circles the point counter-clockwise. This is depicted in Figure 6.24. Singular points with other integer values may be obtained by combining together two or more of the points already described.

The sum of the indices of the singular points of any surface is the same as the Euler characteristic of the surface. Thus a sphere has Euler characteristic $\chi = 2$, and for the flow as depicted in Figure 6.17 there is one sink and one source, each having index 1. The flow depicted in Figure 6.18 has two vortices, each having index 1, again summing to give the Euler characteristic $\chi = 2$. The flows on the torus depicted in Figures 6.19 and 6.20 each confirm the Euler characteristic $\chi = 0$. The fact that the Euler characteristic is a topological property of a surface means that the introduction of a singular point into the flow on a given surface must lead to the appearance of a compensating singular point which will maintain the overall sum of the indices constant. Figure 6.25 depicts what happens if a vortex is introduced into the original flow on

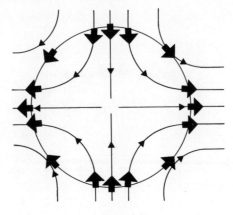

Fig. 6.24

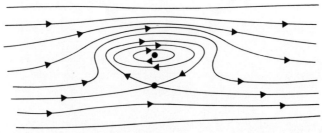

Fig. 6.25

a torus depicted in Figure 6.20. The vortex has index 1, hence a singular point with index -1 has to appear also, in this case a crosspoint similar to that of Figure 6.24.

It is not difficult to show that the sum of the indices of singular points on a surface must be equal to its Euler characteristic. Earlier in this chapter it has been shown that the expression $V - E + F$ is invariant for any map on a given surface, and that this gives the Euler characteristic of the surface. A flow can be constructed on any surface on which a map has been drawn, according to the following three conditions:

1. A source is put at each vertex.
2. A crosspoint is put at the centre of each arc.
3. A sink is put at the centre of each region.

Such a flow constructed on a sphere is depicted in Figure 6.26. There will thus be V sources of index 1, E crosspoints of index -1, and F sinks of index 1, giving a total index of

$$V - E + F,$$

which is the same as the Euler characteristic. Since the introduction of additional singular points into the flow is always compensated so that the index total remains constant, the result is true generally.

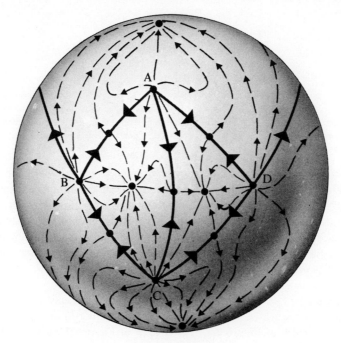

Fig. 6.26

7

Networks

Networks—odd and even vertices—planar and non-planar networks —paths through networks—connected and disconnected networks— trees and co-trees—specifying a network: cutsets and tiesets—travers- ing a network—the Koenigsberg Bridge problem and extensions.

Closely allied to the study of maps on surfaces is the corresponding study of *networks*. A *network* consists of a finite number of vertices linked by a number of arcs. The arcs must be non-intersecting, though two or more may meet at a vertex. No single arc may link directly more than two vertices, and no vertex may be left isolated. Such a figure is also frequently termed a *linear graph*, and the vertices are then usually termed *points* or *nodes*, and the arcs are usually termed *line-segments* or *branches*.

Figures 7.1 to 7.3 depict simple networks. If the number of vertices in a given network is denoted by n and the number of arcs by a, then, for the network of Figure 7.1, $n = 4$, $a = 6$. Figures 7.2 and 7.3 depict five vertices linked by four and seven arcs respectively.

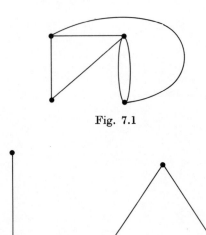

Fig. 7.1

Fig. 7.2

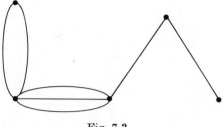

Fig. 7.3

Each vertex is said to have an *order*, this being the number of arc-ends which meet at the vertex. Figure 7.4 depicts a network in which the orders of the various vertices are shown. A vertex whose order is an odd number is termed an *odd vertex*. Similarly, a vertex whose order is an even number is termed an *even vertex*. It is clear that the total

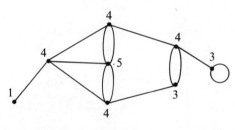

Fig. 7.4

number of arc-ends in any network must be even, since this number must be twice the number of arcs in the network. Further, the number of arc-ends must be equal to the sum of the orders of all the vertices. Since this sum must be even, it follows that the total number of odd vertices must also be even.

If a given network can be mapped on to a simply connected surface (i.e. the plane) in such a way that the non-intersecting property of each and every arc is preserved, then the network is termed *planar*. Figure 7.5. depicts a planar network. This may be mapped on to a sphere as shown in Figure 7.6, or on to a disc as shown in Figure 7.7.

Two *non-planar* networks are depicted in Figures 7.8 and 7.9. Figure 7.8 depicts the familiar problem, well-known to schoolboys, of connecting the mains services of water, gas and electricity to three neighbouring houses. Figure 7.9 depicts what is termed the *complete network on five vertices*, that is, the network obtained when five vertices are joined by the minimum number of arcs in such a way that each

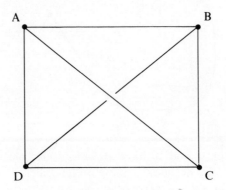

Fig. 7.5

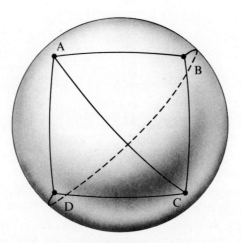

Fig. 7.6

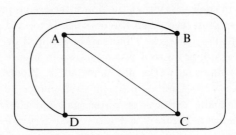

Fig. 7.7

vertex is directly linked to every other vertex. These are the two simplest non-planar networks, and it can be proved that every non-planar network must contain either one or the other of these as a sub-network.

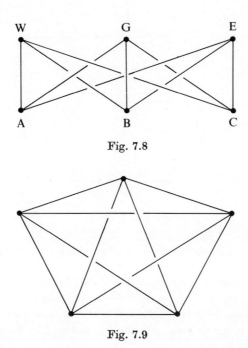

Fig. 7.8

Fig. 7.9

A network is termed *complete* if its vertices are all directly mutually linked by the minimum number of arcs.

A *path* is defined as a sequence of arcs which can be followed continuously without any arc being used more than once. A path is said to *traverse* a network if every arc of the network is included. For example, the four arcs of the network of Figure 7.2 form a single path which traverses the network. A path is said to be *closed* when it starts and finishes at the same vertex; otherwise, it is *open*. A closed path is frequently termed a *circuit*. Figures 7.10 and 7.11 depict an open path and a closed path respectively. In Figure 7.10, the open path is *ABCD*. In Figure 7.11 the closed path is *BCDEFGB*. In neither case is the network traversed, since there are arcs excluded from the paths.

Figure 7.12 depicts an example of a network which cannot be traversed by a single path. The network is comprised of arcs linking three vertices of order one, one vertex of order two, one vertex of order three, and one vertex of order four. Clearly, an even vertex may be the

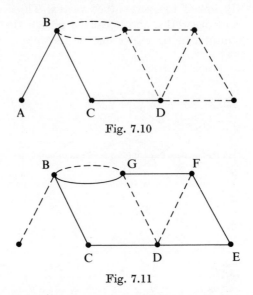

Fig. 7.10

Fig. 7.11

starting and terminating point of a closed path traversing a network. If, however, a path traversing a network is open, then the vertices at which it starts and terminates must be odd. From this it follows that if a network has more than two odd vertices it cannot be traversed by a single path. The network of Figure 7.12 has four odd vertices and cannot be traversed by a single path. The network of Figures 7.10 and 7.11, however, has exactly two odd vertices and can be traversed by a single path. The condition of having exactly two odd vertices is necessary and sufficient for a connected network to be traversed by a single open path.

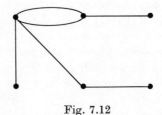

Fig. 7.12

A network is said to be *connected* if every pair of vertices belongs to some path; otherwise it is said to be *disconnected*. The network of Figure 7.12 is connected, but that of Figure 7.13 is disconnected. In Figure 7.13 it will be seen that every individual vertex belongs to some

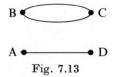

Fig. 7.13

path, but that the vertex-pair AB, for example, does not. If a network is connected, then the total number of its arcs cannot be less than the number of its vertices minus one. This may be written

$$a \geqq n - 1.$$

This is a necessary but not a sufficient condition for a network to be connected. That it is not sufficient can be seen from Figure 7.13, where the network depicted has $a = 3$, $n = 4$. The condition

$$a \geqq n - 1$$

is satisfied, but the network is disconnected.

A connected network, having the number of its arcs exactly one less than the number of its vertices, is called a *tree*. The network of Figure 7.14 is an example of a tree. It will be seen that it has five

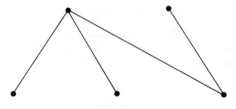

Fig. 7.14

vertices and four arcs. Any connected network can be reduced to a tree by the removal of suitable arcs. Thus, every connected network must include at least one tree. Figure 7.15 depicts a connected network. Removal of those arcs indicated by hatched lines leaves the tree $ABCD$. This tree has four vertices and three arcs.

The arcs removed from a connected network so as to leave a tree form a *co-tree*. Any arc having been so removed forms, when replaced, a closed path in association with one or more of the tree arcs in one way only. Thus, in the network of Figure 7.15, either of the arcs BD forms a closed path with BC and CD, and in no other way. Where a connected network includes more than one tree, there are alternative

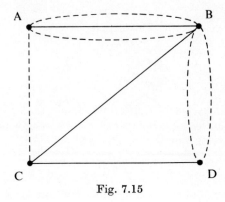

Fig. 7.15

selections of arcs which may be removed so as to leave a tree. This is the case with the network of Figure 7.15, and Figure 7.16 depicts such an alternative selection.

If a network is mapped on to a simply connected surface so that no arcs intersect, except at vertices, then the surface is separated into a number of *bounded regions*. (Such a network must be, by definition, planar.) The number of *independent* bounded regions is given by

$$a-n+1.$$

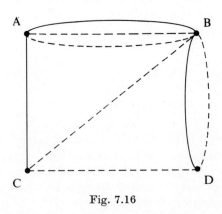

Fig. 7.16

A very simple example is provided by the network having n vertices and comprising a single closed path of n arcs mapped on to a sphere. Clearly, this network divides the surface into two bounded regions. These regions are not, however, independent: once one is defined, the other is automatically defined also. Figure 7.17 depicts a planar network having four vertices and seven arcs. The number of its independent bounded regions is thus $7 - 4 + 1 = 4$. These are the regions R_1 to R_4 shown in the figure. The exterior region is automatically defined.

A planar network may be completely specified either in terms of its independent bounded regions or in terms of paths between its independent vertex-pairs. (A network has $n(n-1)/2$ vertex pairs. It is not necessary in specifying a network for all the paths between each and

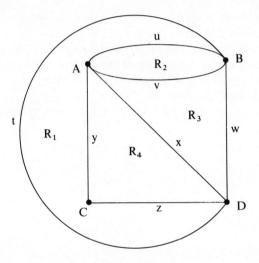

Fig. 7.17

every pair of vertices to be defined. In this sense, the number of *independent* vertex-pairs is $n-1$.) If the network is specified in the former way, then it is said to be defined by *tie-sets*, where a tie-set is a connected sub-set of vertices and arcs, such that exactly two arc-ends meet at each vertex. A tie-set is thus a single closed path of the kind depicted in Figure 7.18. The closed paths which specify the independent

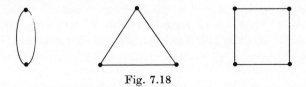

Fig. 7.18

bounded regions do not have to correspond to the paths which form the boundaries of the regions. Thus, in Figure 7.17, the tie-sets specifying the network could be those forming the boundaries of the individual regions, namely

$$tuyz: R_1, \quad uv: R_2, \quad vwx: R_3, \quad xyz: R_4,$$

but they could equally well be

$$tvyz: R_1+R_2, \quad vwzy: R_3+R_4, \quad uwzy: R_2+R_3+R_4, \quad twxyz: \quad R_1+ R_2+R_3,$$

or any other set of four independent closed paths.

If the network is specified by the paths between its independent vertex-pairs, then it is said to be defined by *cut-sets*, where a *cut-set* is a subset of arcs of a connected network such that its removal is exactly sufficient (and no more) to disconnect the original network. In this case, the network may either be separated into distinct subnetworks or a single vertex may be isolated. Figure 7.19 depicts some possible cut-sets. The use of the term 'cut-set' may readily be understood by

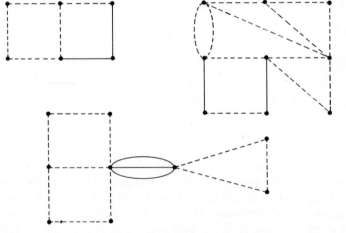

Fig. 7.19

reference to Figure 7.19 and seeing that in each case a cut across the arcs comprising the cut-set would effectively disconnect the network concerned. The network of Figure 7.17 may be defined by cut-sets by, for example, selecting the three independent vertex-pairs A–B, A–C, A–D and specifying all possible paths between the vertices of each individual pair.

If the number of independent bounded regions given by $a-n+1$ is less than the number of independent vertex-pairs, $n-1$, that is, if $a < 2(n-1)$, then tie-set definition of a network may be preferred as requiring a minimum of information for the complete specification of the network. However, if a network is non-planar, tie-set definition is not appropriate, and thus the cut-set approach may be considered to be more general.

If a network is to be traversed by a single path then a first requirement is that it be connected. If there are no odd vertices, then the path must be closed, and it follows that the initial vertex and arc of the path may be selected arbitrarily. If there are exactly two odd vertices, then the path must be open, and it follows that the odd vertices must be selected as the initial and final vertices of the path. For an even number of odd vertices greater than two, it has already been seen that the network cannot be traversed by a single path. Such a network needs at least k paths to traverse it, where $2k$ is the number of odd vertices. This follows because the even vertices can be initially disregarded and, by removal of closed paths, the odd vertices can be reduced to order one. For $2k$ vertices of order one, k separate paths are required, and the closed paths can then be rejoined to various of these as appropriate.

Figure 7.20 (a) and (b) depicts a network with four odd vertices from which three closed paths are removed in order to reduce its odd vertices to order one. First, the even vertex E is disregarded. Then, the two upper closed paths ABA and BCB are removed, as also is the lower closed path DED. Two separate paths AD and BEC remain.

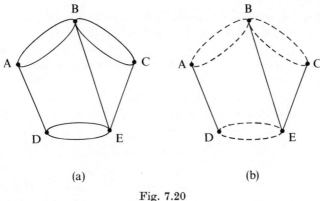

(a)   (b)

Fig. 7.20

The closed paths which have been removed can then be rejoined either to AD or BEC, but because of being closed, the resulting path obtained in each case still terminates at the point where it is rejoined. Thus ABA and DED may be rejoined to AD, giving a total path $ABADED$. There is now no way of linking this path with the remainder of the network since possible connecting arcs have already been used up.

Adding *BCB* to *BEC* gives a second total path *BCBEC*, and this, in addition to *ABADED*, provides the second path by means of which the network is traversed.

One of the best known traditional problems involving the theory of networks is the *Koenigsberg bridge problem*. Figure 7.21 depicts the seven bridges over the divided waters of the River Pregel as they had been built by the eighteenth century. The problem, solved by Euler in 1736, was whether or not it was possible to visit each of the four separated parts of the city whilst crossing each and every bridge once

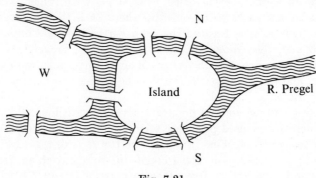

Fig. 7.21

only. The corresponding network is depicted in Figure 7.22. This has four vertices and seven arcs. Examination of the vertices reveals that they are all odd, hence it follows that at least two paths are required to traverse the network, so the whole city could not be toured in the way desired.

The problem may easily be extended. For example, it may be asked where an eighth bridge should be built so that it would be possible

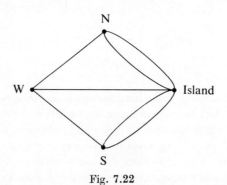

Fig. 7.22

to make the kind of round tour envisaged starting in the North region and finishing on the Island, there still being a requirement to cross each bridge once and once only. Clearly, the effect of such a bridge on the network is to add an arc so as to make two of the odd vertices even. From the previous theory, the vertices at which the path begins and ends must be odd. It follows, therefore, that the eighth bridge should be built from the West region to the South region. If it is now assumed that this bridge has in fact been built, and also a ninth bridge providing a direct link between the North and South regions, the new network

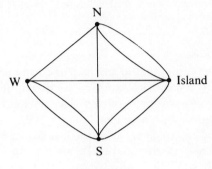

Fig. 7.23

would be as depicted in Figure 7.23. The orders of the vertices are now:

North: 4,   West: 4,   South: 5,   Island: 5.

A round tour, crossing each bridge once and once only, would still be possible, but only if the end points of the path were located in the South region and on the Island respectively.

The study of planar networks has a considerable number of practical applications, particularly in electrical engineering, economics, and sociology. However, topology is more particularly concerned with those properties of networks which distinguish planar networks and the various types of non-planar networks from each other. Such properties belong to the simplest surface on which it is possible to map a given network without intersection of its constituent arcs.

8

The Colouring of Maps

Colouring maps—chromatic number—regular maps—six-colour theorem—general relation to Euler characteristic—five-colour theorem for maps on a sphere.

It is a well accepted fact that printers of maps require only four different colours to ensure that in any map or on any page of an atlas no two regions having a common boundary are given the same colouring. Nevertheless, it has never been proved that the existence of a planar map requiring five colours is an impossibility.

It is clear that three colours alone are insufficient to meet the needs of the printers. The map depicted in Figure 8.1, for example, demonstrates this fact. It would be impossible to colour the region in the centre with one of the three colours already used whilst still meeting the requirement that there should always be different colourings on two sides of any boundary. It will be proved later in this chapter that five colours are sufficient for the colouring of any map on a plane surface or on a sphere satisfying the printers' requirement.

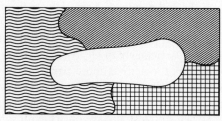

Fig. 8.1

The least number of colours required to colour a map on any given surface is termed its *chromatic number*, and it is a topological property of the surface. Thus, for a disc and for a sphere, it is assumed that the chromatic number c is equal to four, though this has never been proved. A map drawn on a finite plane surface is equivalent to the corresponding map on a sphere. This follows from the fact that a disc may be cut out of the centre of any one region of a map on a sphere, as

70

depicted in Figure 8.2, and the resulting surface may then be deformed so as to lie in a plane. The region from which the disc has been removed then forms the surrounding region of the map on the plane. Similarly, if the surrounding area of any map on a plane is counted as a region, then this region may be joined up at all its outer edges and the plane thus transformed into a sphere.

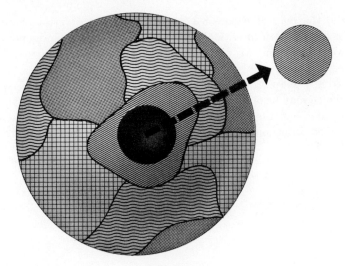

Fig. 8.2

Before proceeding with a detailed consideration of the colouring of maps, it is necessary first to refer to the definition of a map given at the beginning of Chapter 6. This definition states that a map consists of a number of vertices linked together by non-intersecting arcs in such a way that simply-connected regions are defined by the area. Figure 8.3 depicts such a map, but, clearly, this does not correspond to any map which could be found on the page of an atlas. Not only are vertices C, D and F superfluous, but, in addition, the pair of arcs EF, EG play no part in defining any region, since the areas immediately above and immediately below the arc-pair belong to one and the same region of the map.

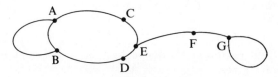

Fig. 8.3

A map is said to be *regular* if, in addition to complying with the requirements of the definition given above, it also satisfies the following three conditions:

1. No vertex is of order less than three.
2. Each arc joins two distinct vertices.
3. Each arc separates two distinct regions.

Clearly, the map of Figure 8.3 is not regular since vertices C, D, F do not comply with condition (1), the loop from vertex G does not comply with condition (2), and arcs EF, FG do not comply with condition (3).

To any map on a closed surface there corresponds some regular map which may be obtained from it. (The same applies to any map on a finite plane, but it is sufficient here to consider the case where the map is assumed drawn on a closed surface.) Figure 8.4, for example, shows a

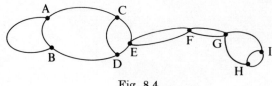

Fig. 8.4

regular map obtained from the map of Figure 8.3. In order to arrive at a regular map, it has been necessary to add the two vertices H, I together with several arcs, thus increasing the total number of distinct regions from four to eight. The map of Figure 8.4 does correspond to one which might possibly be found on the page of an atlas. In fact, with the exception of the cases where one region is entirely surrounded by another, geographical and similar maps satisfy the conditions for regular maps on a finite plane or on a sphere. Clearly, when one region is entirely surrounded by another, there is no special problem of colouring involved since any colour different from that of the surrounding region may be used.

If a regular map having F regions, E arcs, and V vertices is defined on a surface of Euler characteristic $\chi > 0$, where from Chapter 6, $\chi = V - E + F$, then $V - E + F > 0$, and thus

$$6V - 6E + 6F > 0.$$

It follows from condition (1) above that $2E \geq 3V$, and hence substitution for $6V$ gives

$$6F - 2E > 0.$$

This may be expressed, using condition (3), in the form

$$6\Sigma F_i - \Sigma i F_i > 0$$

where F_i is the number of regions with boundaries formed of exactly i arcs, and where, clearly by condition (2), i is greater than one. This inequality may be written in the form

$$\Sigma(6-i)F_i > 0,$$

from which it is seen that some i less than six must exist. Thus, any regular map on a surface of Euler characteristic greater than zero must have at least one region bounded by fewer than six arcs.

It is now possible to prove by induction the result that any regular map on a surface of Euler characteristic greater than zero requires at most six colours if no adjoining regions are to be coloured the same. If $F \leq 6$, then the required result is immediate. For general F, it is first supposed that the result holds for some F', and then the map with $F' + 1$ regions is considered. It has already been shown that at least one region must have a boundary consisting of less than six arcs. If this region is contracted to a point, there are four possibilities, and these are depicted in Figures 8.5 to 8.8. In each of the cases depicted the contracted map, having F' regions, is by assumption colourable with six colours. So, when the contracted region is restored, there will be a colour available for it without the total of six colours being exceeded. Thus, if the result holds for $F = F'$, then it holds also for $F = F' + 1$. Now, since maps for which $F \leq 6$ can be coloured with not more than six colours, the general result that any regular map on a surface, for which $\chi > 0$, requires at most six colours follows by induction. This general result is termed the *six colour theorem*.

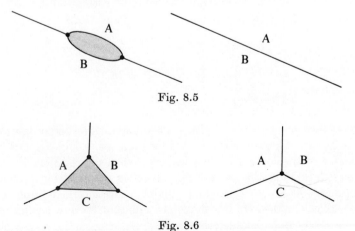

Fig. 8.5

Fig. 8.6

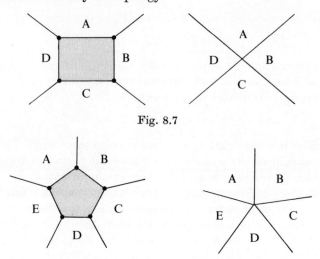

Fig. 8.7

Fig. 8.8

A more general theorem states that any regular map on a surface of Euler characteristic χ can be coloured by at most γ colours, where γ satisfies the inequality

$$\gamma F > 6(F - \chi) \tag{1}$$

for all $F > \gamma$. The validity of this theorem may be demonstrated by an inductive proof similar to that used in substantiating the six colour theorem.

First, it is assumed for some given surface of Euler characteristic χ, that γ, satisfying the inequality (1), is such that all regular maps having $F \leq F'$ on the surface may be coloured with at most γ colours. In particular this is trivially true for $F \leq \gamma$. Now, since

$$\chi = V - E + F, \text{ and } 2E \geq 3V,$$

it follows that

$$6(F - \chi) = 6(E - V) \geq 2E,$$

whence, from (1),

$$\gamma F > 2E.$$

Thus, at least one region must have a boundary comprised of less than γ arcs. If such a region is contracted into a point, χ is unchanged and the map is still colourable with γ colours. When the contracted region is restored, there is still, at worst, a γ'th colour available for it. Hence, by a process of inductive reasoning, analogous to that used in the proof of the six colour theorem, γ colours are sufficient for any regular map on a surface of Euler characteristic χ.

The problem remaining is the determination of the smallest integer γ satisfying inequality (1) when $F > \gamma$. Let this smallest integer be denoted by $\tilde{\gamma}$. Now six colour theorem gives

$$\gamma = 6 \text{ when } \chi = 2 \text{ and when } \chi = 1.$$

If the inequality (1) is expressed in the form

$$\gamma > 6\left(1 - \frac{\chi}{F}\right), \tag{2}$$

it can immediately be seen that its right-hand side approaches six from below with increasing F when $\chi = 2$ or 1. Thus (2) gives

$$\tilde{\gamma} = 6 \text{ for } \chi = 2 \text{ and } F > 12,$$

and also gives

$$\tilde{\gamma} = 6 \text{ for } \chi = 1 \text{ and } F > 6.$$

These values of $\tilde{\gamma}$ equate with the value six of the six colour theorem.

When $\chi = 0$, the right-hand side of inequality (2) is exactly equal to six, and hence the expression gives

$$\tilde{\gamma} = 7 \text{ for } \chi = 0.$$

For $\chi < 0$, the smallest admissible $\gamma + 1$ may be substituted for F in (2) to give

$$\gamma > 6\left(1 - \frac{\chi}{\gamma + 1}\right).$$

This may be rearranged as

$$\gamma(\gamma + 1) > 6\gamma + 6 - 6\chi,$$

whence

$$\gamma^2 - 5\gamma > 6 - 6\chi,$$

$$(\gamma - \tfrac{5}{2})^2 > \tfrac{49}{4} - 6\chi$$

and thus

$$\gamma > \tfrac{5}{2} + \tfrac{1}{2}\sqrt{(49 - 24\chi)}.$$

Using square brackets to denote the largest integer in a given expression gives

$$\tilde{\gamma} = [\tfrac{7}{2} + \tfrac{1}{2}\sqrt{(49 - 24\chi)}]. \tag{3}$$

(The positive root is intended to be taken.) Expression (3) gives a minimum value for the number of colours which are required at most for the colouring of a regular map on a surface of Euler characteristic less than zero. The table following, gives the values of $\tilde{\gamma}$ obtained from expression (3) for values of χ from two to minus twelve. (The values for

$\chi = 2, 1$ and 0 have been given in brackets because, though these are values which from other considerations would be expected, they are not strictly applicable because of the original assumption that $\chi < 0$, made when determining the expression (3) for $\tilde{\gamma}$.) It can be seen from the table that, for example, any map on a Klein bottle requires at most seven colours, and any map on a two-fold torus requires at most eight colours.

χ	$\tilde{\gamma}$
(2)	(4)
(1)	(6)
(0)	(7)
-1	7
-2	8
-3	9
-4	9
-5	10
-6	10
-7	10
-8	11
-9	11
-10	12
-11	12
-12	12

For regular maps on a sphere, and hence also on a plane, it has been proved that the maximum number of colours required is six. It can be demonstrated that this maximum can be reduced by one to five in the following way.

It has already been established that in any regular map on a sphere some region must have a boundary consisting of fewer than six arcs. Clearly, any region with two, three, or four arcs forming its boundary must be colourable with a fifth colour. Figure 8.9 depicts a region having its boundary made up of exactly five arcs. Now, some pair of the regions A, B, C, D, E must have no common boundary. This is demonstrable as follows. If, for example, regions A and C adjoin, then region B must be isolated from regions D and E. Region B can therefore be coloured with the same colour as either of the two regions D and E. Since this holds generally, a fifth colour is always available for the region having its boundary made up of exactly five arcs. If one arc is removed from a region bounded by two, three, or four arcs, so as to make the region coalesce with an adjoining one, then a map with $F-1$ regions results where F is the number of regions of the original map. If this map can be coloured with five colours, then so can the original map. If two arcs are removed from a region bounded by five arcs in such a way as to make the region coalesce with a separated pair of regions, then a map with $F-2$ regions results, and, if this can be coloured with five colours, then so can the original map. The number of regions can therefore be

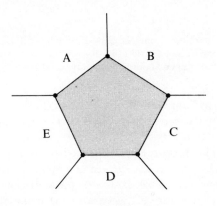

Fig. 8.9

successively reduced by the removal of arcs so as to lead to a sequence of maps having fewer and fewer regions, and, if these maps can be coloured with five colours, then so can the original map. Eventually, a map is arrived at having less than six regions in all, and, since this can clearly be coloured with at most five colours, the original map does not require more than five. This result establishes what is termed *Heawood's five colour theorem*.

It should be remembered that, in the cases of the six and five colour theorems and the derivation of the expression for γ (known as *Heawood's theorem*), what has in fact been demonstrated is the sufficiency of a certain number of colours for the colouring of any regular map on a given surface of known Euler characteristic. The *necessity* for any specific number of colours in any particular case has still to be justified. Figure 8.1 establishes the necessity for at least four colours for maps on a surface of Euler characteristic $\chi = 2$. In certain other cases, notably those for surfaces with Euler characteristic $\chi = 1$ or 0 or χ an even negative integer, the value obtained from expression (5) is necessary as well as sufficient. In these cases, therefore, the calculated values for γ are also the chromatic numbers of the surfaces concerned. Further reference will be made to this topic following the initial discussion of plane diagrams in Chapter 11.

9

The Jordan Curve Theorem

Separating properties of simple closed curves—difficulty of general proof—definition of inside and outside—polygonal paths in a plane—proof of a Jordan curve theorem for polygonal paths.

At the beginning of Chapter 1, it was assumed that one of the properties of a triangle is that it separates a plane surface into an area inside its perimeter and an area outside its perimeter. Again, at the beginning of Chapter 4, a similar property was assumed for a non-self-intersecting continuous closed curve on the surface of a sphere. These assumptions are intuitively very reasonable. It would appear obvious that the closed curve C, depicted in Figure 9.1, separates the plane of the paper into a set of points, such as A, lying inside the curve, and a set of points, such as B lying outside the curve (neglecting for the moment the set of points in the curve C itself).

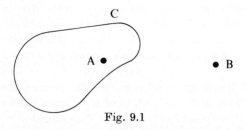

Fig. 9.1

Figure 9.2 also depicts a continuous non-self-intersecting closed curve on a plane surface. However, it is now not so immediately obvious that points A and B lie respectively inside and outside the curve. Thus, the more complex the curve being investigated, the greater is the need for some sort of formal test to determine for any given point whether it lies inside or outside the curve. Further, the terms 'inside' and 'outside' have been used here and earlier in a purely intuitive kind of way, and there is also a need for defining exactly what these two terms mean.

Any curve *homeomorphic* (i.e. topologically equivalent) to a circle is termed a *Jordan curve*. Thus, the curves of Figures 9.1 and 9.2 are both Jordan curves. The *Jordan curve theorem* states that on a plane or on the surface of a sphere a Jordan curve separates the surface into two regions having no point in common and having the curve as a common boundary. Surprisingly perhaps, a general proof of this theorem is not particularly easy to present. The special case when the curve defines a closed polygonal path will be considered here.

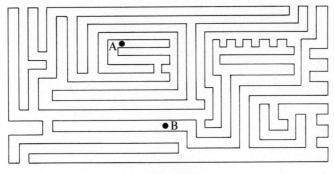

Fig. 9.2

First, it is necessary to define a *polygonal path*. This is defined as n straight line segments joining a sequence of n distinct points in a plane in such a way that no two line segments intersect except possibly at their end points, and each line segment joins two points uniquely. The number n is assumed to be finite. The straight line segments form the *sides* of the polygon whose perimeter is the given polygonal path.

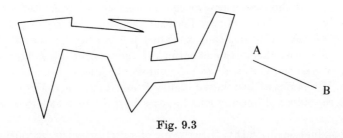

Fig. 9.3

Since n is finite, it must be possible to choose some straight line segment AB in the plane which is not parallel to any of the sides of the polygon. Figure 9.3 depicts a polygonal path and a line segment AB not parallel to any side of the polygon defined. The set of points of the plane not on the polygonal path itself is now separated into two disjoint subsets, the subset to which any given point of the set belongs being determined according to whether a ray from the point in a direction parallel to AB intersects the polygonal path an even or an odd number of times. If these subsets are labelled S and T respectively, then it is clear that, for example, points P_1, P_2, P_4, depicted in Figure 9.4. belong to subset S, whilst points P_3, P_5 belong to subset T.

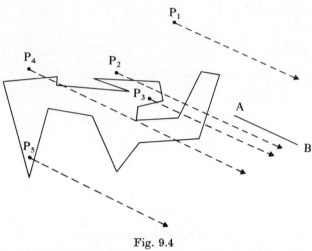

Fig. 9.4

It is now assumed that a variable point P moves continuously along the length of some line segment not intersecting the polygonal path, and not parallel to AB. As P moves, the number of intersections of its ray parallel to AB with the polygonal path changes only when the ray crosses a vertex. Thus the number of intersections always changes by two, or, in the case where more than one vertex is crossed simultaneously, by a multiple of two. For example, as a point P moves along the line segment CD, depicted in Figure 9.5, its ray has initially no intersections with the polygonal path. As P moves from a to b, however, its ray crosses the vertex W and the number of intersections increases to two. It remains at two until P moves from c to d, when Y is crossed by the ray and the intersections increase by a further two to a total of four.

It should be noted that the crossing of a vertex by the ray does not necessarily lead to a change in the number of intersections with the polygonal path. For example, as P moves from d to e two vertices are crossed, but the number of intersections remains unchanged at four. As the point moves from e to f, vertex X is crossed and the number of intersections decreases, again by two, but, as P moves from f to g, vertices V and Z are crossed simultaneously, and there is an increase of four in the number of intersections bringing the total up to six. For

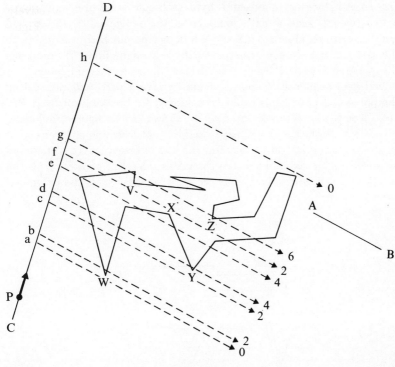

Fig. 9.5

each position of P on CD, the number of intersections is indicated in
Figure 9.5 at the arrow-head of the appropriate ray. Eventually, P
reaches h, by which time the ray has entirely cleared the polygonal
path, and the number of intersections has reduced to zero and remains
zero thereafter. Similar considerations apply to a point moving along
the line segment EF depicted in Figure 9.6.

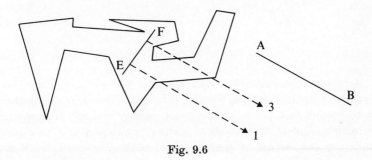

Fig. 9.6

It now follows that if any point belonging to subset S is joined to any point belonging to subset T by a path consisting of a sequence of line segments, such a path must cross the polygonal path an odd number of times. Figures 9.7 and 9.8 depict points P, P' belonging to the same subset. In each case, even if the straight line PP' intersects the polygonal path it can be seen that the points could have been joined by a sequence of line segments forming a path not intersecting the polygonal path. This could be achieved by breaking off from PP' at a, just before the first intersection shown in the figures, following round close to the polygonal path itself but not touching or intersecting it, and rejoining PP' at b just after the last intersection. It is therefore possible in each case to identify subset S as the set of points *outside* the polygonal path, and subset T as the set of points *inside* the polygonal path.

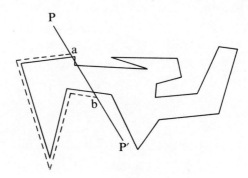

Fig. 9.7

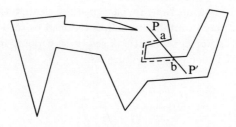

Fig. 9.8

An alternative method of separating the set of points of the plane into the two distinct subsets S and T is to define the *order* of any point P with respect to a polygonal path not passing through P as the net number of complete revolutions of a straight line joining P to a point completing one circuit of the polygonal path. For example, if the point P', depicted in Figure 9.9, makes a circuit of the polygonal path in a counter-clockwise direction as shown in the figure, then the total number of revolutions completed by the straight line P_1P' is one. In the same way, the total number completed by P_2P' is zero. The subset S may now be defined as the subset of all points of the plane of even order, and the subset T as the subset of all points of the plane of odd order. These two subsets correspond exactly to those defined in terms of the intersections of rays with the polygonal path.

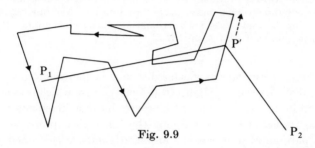

Fig. 9.9

The ease with which the Jordan curve theorem may be justified in the case of a polygonal path is, of course, due to the fact that the finite number of sides of the polygon defined permits the definition of a direction not parallel to any side. Clearly, once curved paths are permitted, it may no longer be possible to define a direction which is in no instance tangential to any curve. For example, for a circular path in a plane, there is no straight line in the plane which is not parallel to some tangent to the circle. The general case is, however, of considerable importance in topology. It also provides a typical example of the fact that it is not always simple to *prove* what is in fact true even when it appears intuitively obvious.

10

Fixed Point Theorems

Rotating a disc: fixed point at centre—contrast with annulus—continuous transformation of disc to itself—fixed point principle—simple one-dimensional case—proof based on labelling line segments—two-dimensional case with triangles—Sperner's lemma—three-dimensional case with tetrahedra.

One of the simplest ways in which a disc may be mapped to itself is by a rotation about its centre. Figure 10.1 depicts a disc which is assumed to be rotated in its own plane about its centre by some fixed angle ϕ. This is a rigid transformation preserving the topological properties of the disc. In particular, it is one–one and continuous so that each point x of the disc is mapped to some unique point x', which is the image of x

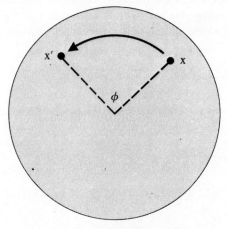

Fig. 10.1

and of no other point, and neighbourhoods are preserved so that points which are 'near' remain 'near'. (A more precise discussion of this concept of *continuity* will be given in Chapter 13.)

It is clear that for the rotation through ϕ depicted in Figure 10.1 there is one point and, for ϕ not an integer multiple of 2π, one point only which maps to itself, namely the centre of the disc. On the other hand, for a similar rotation of an annulus, depicted in Figure 10.2, it is clear that, unless ϕ is an integer multiple of 2π, there is no point which is mapped to itself.

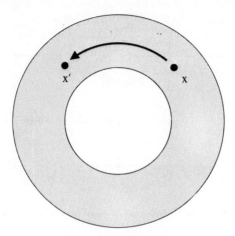

Fig. 10.2

The example given by the rotation of a disc is a simple and very obvious illustration of *Brouwer's fixed point theorem* in two dimensions. However, the general case of this theorem is by no means so obvious. The theorem states that, for any continuous transformation of a disc to itself, there is at least one point which is mapped to itself. Since this is in fact a topological property, it holds equally for any region homeomorphic to a disc. For example, if the wind blows over the surface of a pool of oil in the road, then, provided the surface of the oil is moved about the same overall area and is not broken in any way, there is at least one point at any one time where the oil is in exactly the same place as it was originally before the wind began to blow.

A similar fixed point theorem may be stated in one dimension. In this case instead of a disc, a line segment or interval is considered, and the theorem states that if an interval is continuously transformed to itself there is at least one point of the interval which remains fixed.

One way of arriving at proofs of the fixed point theorems is to start from a consideration of dividing up a triangle into small triangles (in the two-dimensional case) or a line segment into small line segments (in the one-dimensional case) according to certain rules which also include adopting a specific system of numbering of vertices and endpoints of segments. Figure 10.3 depicts a line segment AB which is divided into small segments by arbitrarily selecting a number of its points. These points and the original end-points *A* and *B* are then labelled 0 or 1 in an arbitrary manner, and a cross is put on each side of every 0 within the original line segment *AB*, and on the 'inside' only of any 0 which may appear at the end-points of *AB*. Figure 10.4 depicts several possible ways of dividing up *AB* in this way.

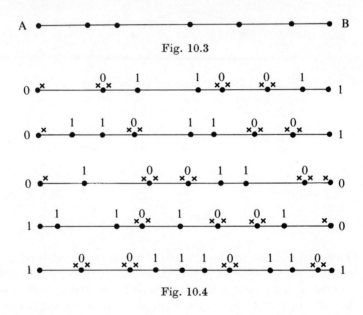

Fig. 10.3

Fig. 10.4

A *complete segment* is now defined as a small segment having a 0 at one end and a 1 at the other. Thus, for the top example depicted in Figure 10.4, there are altogether seven small segments whose end-points are labelled successively

$$00, \mathbf{01}, 11, \mathbf{10}, 00, \mathbf{01}, 11,$$

and for the ·bottom example there are ten small segments labelled successively

$$\mathbf{10}, 00, \mathbf{01}, 11, 11, \mathbf{10}, \mathbf{01}, 11, \mathbf{10}, \mathbf{01}.$$

In the former case there are three complete segments (shown in bold face), and in the latter case there are six. It is clear that each complete segment has one cross, whereas any other small segment has either no cross or two crosses.

If *n* is the total number of complete segments in any particular decomposition of a given line segment, then the total number of crosses is *n* + some even positive integer. However, the total number of crosses is also given by the number of 0's at the end-points of the original line segment, which each have one cross, plus an even positive integer representing twice the number of 'internal' 0's (which will each have two crosses). Since these two totals must be equal it follows that if the number of 0's at the end-points of the original line segment is odd, that is, if the end-points of the original line segments are labelled 0 and 1, then the number of complete segments must also be odd. This in turn means that there must be at least one complete segment however the original line segment (labelled **01** or **10**) is divided up.

If the original line segment, assumed to be labelled **01**, is now continuously transformed to itself, then, subsequent to this transformation, points whose distance from the end-point labelled 0 has not decreased may be labelled 0, and points whose distance from the end-point labelled 1 has not decreased may be labelled 1. The earlier result now means that there must be some segment, which may be chosen as arbitrarily small as desired (since the process of labelling points may be continued indefinitely) such that at least one of its points has not decreased in distance from the end-point of the original line segment labelled 0 and at least one of its points has not decreased in distance from the end-point labelled 1. In the limit, this arbitrarily small complete segment tends to a single point, which is then the fixed point whose existence the argument has been seeking to justify.

Figure 10.5 depicts a triangle which has been arbitrarily divided up into small triangles. The vertices of the original triangles and the

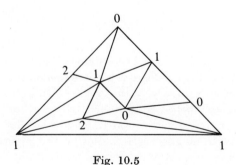

Fig. 10.5

small triangles have then been arbitrarily labelled 0, 1, or 2. Crosses are shown in the figure and these have been placed inside the original triangle against every line segment labelled 01. Thus a line segment labelled 01 on the boundary of the original triangle has one cross (on the inside), whereas a line segment labelled 01 inside the original triangle has two crosses (one on either side). A *complete triangle* is now defined as a small triangle whose vertices are labelled 012. Thus, there are three complete triangles in Figure 10.5.

If n is the total number of complete triangles in any particular decomposition of an original triangle into small triangles, then the number of crosses will be $n +$ some even positive integer. This follows since each complete triangle has one cross, whereas all other small triangles have either no cross or two crosses. (The possible cases are depicted in Figure 10.6.) However, the total number of crosses are also given by the number of line segments labelled 01 on the boundary of the original triangle, which each have one cross, plus an even positive integer representing twice the number of internal line segments labelled 01 (which will each have two crosses). Since these two totals must be equal it follows that if the number of line segments labelled 01 on the boundary of the original triangle is odd, then the number of complete triangles must also be odd.

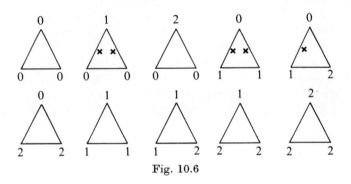

Fig. 10.6

A special case of the result just obtained is provided if the original triangle is labelled 012, and there is a restriction on the labelling of the line segments on its boundary, so that vertices lying on the original side labelled 01 may be labelled either 0 or 1 only, vertices on 02 may be labelled 0 or 2 only, and vertices on 12 may be labelled 1 or 2 only. A decomposition with numbering according to this restricted system is depicted in Figure 10.7. From the general result, it now follows that for such a special case there is always at least one complete triangle. This special form of the theorem is known as *Sperner's lemma*, which may be used to prove the two-dimensional fixed-point theorem in a manner analogous to the proof for the one-dimensional case.

In three dimensions, a decomposition of a tetrahedron is considered, with numbering of vertices using 0, 1, 2 and 3. A complete tetrahedron is then defined as a small tetrahedron having its vertices labelled 0123,

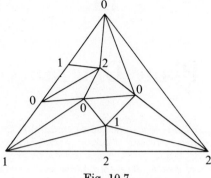

Fig. 10.7

and a corresponding three-dimensional fixed-point theorem may then
be proved. In more than four dimensions, similar arguments may be
used, though the decompositions are much less easily visualised.

11

Plane Diagrams

Definition of manifold—construction of manifolds from rectangle—
plane diagram representations of sphere, torus, Moebius band, etc.—
the real projective plane—Euler characteristic from plane diagrams—
seven colour theorem on a torus—symbolic representation of surfaces
—indication of open and closed surfaces—orientability.

A two dimensional *manifold* is a connected surface having the property
that, given any point P on the surface all the points near to P on the
surface together with the point P itself form a set of points which is
topologically equivalent to a disc. The set of points near to P is termed
a *neighbourhood* of P. A sphere, a torus, a Moebius band and a Klein
bottle all provide examples of surfaces which are also manifolds. A
sphere with a spike sticking out of it, as depicted in Figure 11.1, is not,

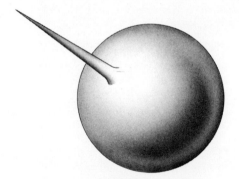

Fig. 11.1

however, a manifold since points on the spike do not have neighbourhoods fulfilling the requirements for all points on a manifold and, in
particular, there is a dimensional change at the spike. (The spike must
not be thought of as a 'thin cone', otherwise the sphere with the spike
is still a manifold.) It should be noted also that a manifold is not
necessarily a closed surface; neither is it necessarily two-sided.

Certain manifolds may be identified by specifying the way in
which the sides of a rectangle are to be joined together. Figure 11.2

depicts a cylinder with curved boundaries x and y. The curved surface of the cylinder may be cut in a direction perpendicular to its boundaries as depicted in Figure 11.3, where such a cut is shown from A to B. It can then be opened up to form the rectangle of Figure 11.4. The method adopted for labelling the vertices of the rectangle ensures that there can be no doubt as to the correct way in which the cylinder is to be reconstituted. This is shown by the use of the identical letter a for the two sides of the rectangle which are to be joined up, together with an indication by means of arrow-heads that there is to be no twisting before the two sides a are joined. A diagram such as that of Figure 11.4

Fig. 11.2

Fig. 11.3

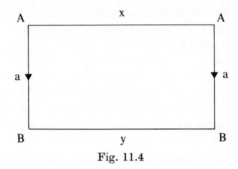

Fig. 11.4

is termed a *plane diagram*. Plane diagrams do not necessarily have to be rectangular, but in this chapter only rectangular plane diagrams will be discussed.

Figure 11.5 depicts a way in which a sphere may be cut in order to give the plane diagram of Figure 11.6. Clearly, some deformation other than simple unfolding is needed in order to obtain the rectangle from the sphere, and also to reconstitute the sphere from the rectangle.

In Figures 11.3 and 11.4, x and y are used for boundaries of the cylinder which become sides of the corresponding rectangular plane diagram, whilst a is used for the sides of the cut which have subsequently to be rejoined in order to restore the original surface. Again, in Figures 11.5 and 11.6, a and b are used for sides of cuts, no other letters being needed since a sphere is a closed surface. In general, small letters from the latter part of the alphabet will be reserved for sides of plane diagrams representing original boundaries of surfaces, whilst

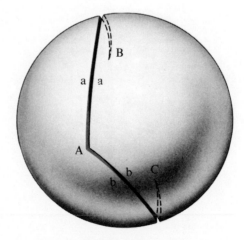

Fig. 11.5

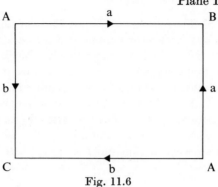

Fig. 11.6

small letters from the beginning of the alphabet will be reserved for sides resulting from cuts.

Figure 11.7 depicts a one-fold torus together with cuts needed to obtain its corresponding plane diagram, depicted in Figure 11.8. Here again, the surface is closed, hence the small letters used come entirely from the early part of the alphabet. The fact that the surface is closed is also inferable from the plane diagram because the four vertices of the rectangle are all labelled with the same letter. Not all closed surfaces, however, give rise to rectangular plane diagrams having all four vertices

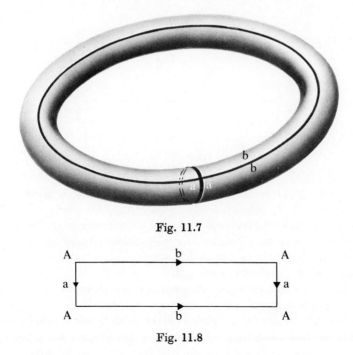

Fig. 11.7

Fig. 11.8

identically labelled. The rectangle of Figure 11.6, for example, identifies a closed surface, but three letters are needed for its vertices.

Figures 11.9 and 11.10 depict respectively a Moebius band and its corresponding plane diagram. Only one cut is needed in this case, but a subsequent twist must be made before the manifold can be represented in a plane. Although the Moebius band has only one boundary, it is necessary to denote this in its plane diagram by separate letters x and y, since the sides of the plane diagram denoted by those letters are not joined together when the manifold is reconstituted. (In fact, they are joined end to end.)

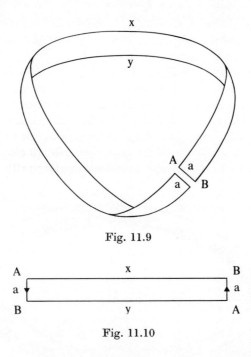

Fig. 11.9

Fig. 11.10

Figure 11.11 is the plane diagram representing a Klein bottle. In this case, it is not possible to reconstitute the surface by manipulation in three-dimensional space. (This can easily be verified by practical experiment.) The fact that the surface ultimately obtained is closed is immediately seen from the use of the same letter A as the label for all four vertices of the rectangle.

Reference to the plane diagram for the Moebius band (Figure 11.10) reveals a special relation between that diagram and that for the Klein bottle. If arrows are placed on the sides labelled x and y of the former so that consistency of direction is maintained when these are

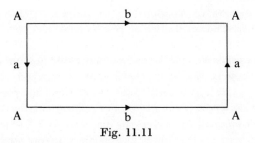

Fig. 11.11

joined end to end to form to manifold, then the resulting diagram with these additional arrows is that of Figure 11.12. Comparison of Figures 11.11 and 11.12 now shows that a Klein bottle can be regarded as a Moebius band with its continuous boundary divided into two parts which are then joined up in a particular way which, in fact, requires four-dimensional space for its completion.

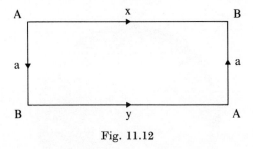

Fig. 11.12

In order to reconstitute the Klein bottle, the opposite sides of its rectangular plane diagram are joined, one pair in the same sense and the other pair in the opposite sense, all vertices coming together so as to form a closed surface. Figure 11.13 depicts a plane diagram where again opposite sides of the rectangle are to be joined together, but this

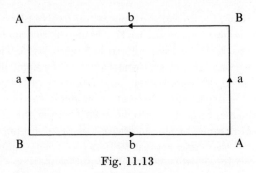

Fig. 11.13

time both pairs are to be joined in the opposite sense leaving two vertices distinct. The surface which results is termed the *real projective plane*.

The real projective plane may be represented in a number of ways. For example, the points of a sphere may be mapped to points in a tangential plane as depicted in Figure 11.14. Here, the plane is tangential to the sphere at the point T. Points P_1, P_2 are mapped to the single point P' in the plane by extending the diameter P_1OP_2 so as to meet the plane at P'. Every great circle of the sphere is mapped to a line in the tangential plane with the exception of the equator defined by the plane through the centre of the sphere parallel to the tangential plane. In order to allow for the mapping of the equator, a *line at infinity* is added to the Euclidean tangential plane, each point of this line representing a pair of diametrically opposite points of the equator. This extended plane is the real projective plane. Any straight line through a given point P in ordinary three-dimensional Euclidean space is a point of the real projective plane.

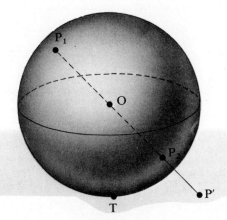

Fig. 11.14

Another way of representing the real projective plane is to project each point of a hemisphere on to the plane of its equator by a line perpendicular to this plane as depicted in Figure 11.15. There is now a one–one correspondence between the points of the hemisphere and the points on and inside a circle. If each pair of diametrically opposite points on the circumference of this circle are now made to coincide, a representation of the real projective plane results. Comparison of Figure 11.13 with the circumference of the equatorial plane of Figure 11.15 shows that essentially the same representation has been obtained, since a circle and a rectangle are topologically equivalent.

The real projective plane may also be constituted from a Moebius band and a disc. The boundary of a Moebius band is a closed curve topologically equivalent to a circle. It can therefore be imagined attached by its boundary to the boundary of the disc so as to form a closed surface. This resulting closed surface is again the real projective plane. Thus a Moebius band may be thought of as the real projective

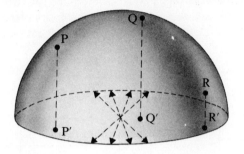

Fig. 11.15

plane with a disc cut out of it. Like a Moebius band and a Klein bottle the real projective plane is one-sided. The deformations needed to produce the real projective plane cannot be performed in ordinary three-dimensional Euclidean space.

The Euler characteristic of a manifold may easily be determined from its plane diagram using the already established expression

$$\chi = V - E + F$$

For example, the plane diagram for the cylinder (Figure 11.4) has two distinct vertices only, namely A and B, since the two upper and the two lower pairs of vertices are to be identified respectively with each other. Further, this rectangular plane diagram has only three distinct sides (edges), namely a, x and y, since the two sides a are to be identified. There is, of course one face. Hence for the cylinder, $\chi = 2 - 3 + 1 = 0$.

The plane diagram for the sphere, depicted in Figure 11.6, has three vertices, two distinct sides, and, of course, one face, giving $\chi = 3 - 2 + 1 = 2$. Again, the plane diagram for the torus (Figure 11.8) gives $\chi = 1 - 2 + 1 = 0$ and that for the Moebius band (Figures 11.10) gives $\chi = 2 - 3 + 1 = 0$. (In this case, the sides x and y are regarded as distinct, since they do not become identified, that is 'sewn up', when the Moebius band is re-constituted, but only joined end to end.) The plane diagram for the Klein bottle (Figure 11.11) gives $\chi = 1 - 2 + 1 = 0$, and that for the real projective plane (Figure 11.13) gives $\chi = 2 - 2 + 1 = 1$. This last provides the first example encountered so far of a manifold whose Euler characteristic is an odd number. The possibility of such surfaces was envisaged, however, when in Chapter 8 the table of upper bounds of the number of colours required for maps upon various surfaces was calculated.

Reference to the table of Chapter 8, p. 76 gives an upper bound $\tilde{\gamma}$ of the number of colours required for any map on the surface of a torus as seven. It is now relatively simple to determine the chromatic number for this surface by reference to a map drawn on its plane diagram as is depicted in Figure 11.16. In viewing this map, it must be remembered

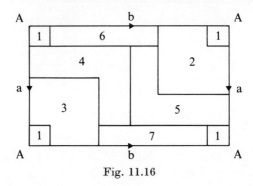

Fig. 11.16

that edge to edge adjacencies apply because of the identification of the two upper and the two lower sides respectively of the rectangle. This map clearly shows a case where seven colours are essential in order to ensure that no two regions having a common boundary are coloured with the same colour. Since an upper bound for γ is seven, and since at least one case can be found where seven colours are required, it follows that seven colours are both necessary and sufficient for maps on the surface of a torus; that is, the chromatic number of the surface is seven. This result is often known as the *seven colour theorem*.

A symbolic representation of manifolds based upon the sides of the corresponding plane diagrams is often used. This method depends not only on the identification of the sides, but also on their directional sense. To take account of directional sense a reference orientation is assigned to the perimeter as a whole (i.e. clockwise or counter-clockwise) and a positive or negative sense is recorded for each edge accordingly as its arrow agrees or disagrees with this reference.

Figure 11.6 represented the plane diagram for the surface of a sphere. Commencing at the top left-hand of the diagram and assigning a clockwise orientation to the perimeter of the rectangle yields the following order of sides:

$$+a-a+b-b.$$

Clearly, any vertex of the rectangle may be taken as the starting point and, since either a clockwise or a counter-clockwise overall orientation may be assigned, this immediately yields three alternative but equivalent symbolic representations:

$$-a+b-b+a,$$

$$+b-b+a-a,$$

$$-b+a-a+b.$$

An alternative plane diagram representing the surface of a sphere is depicted in Figure 11.17. The only relevant difference is that whilst the b sides still 'converge' according to the sense of their arrow-heads, the a sides now 'diverge'. This gives rise to the further symbolic representations:

$$-a+a+b-b,$$

$$+a+b-b-a,$$

$$+b-b-a+a,$$

$$-b-a+a+b.$$

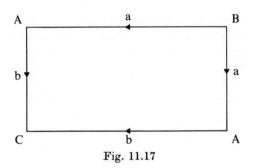

Fig. 11.17

It is clear that the symbolic representation of a sphere requires two letters only, each letter appearing once with positive sign and once with negative sign, and that any cyclic permutation preserving the adjacency of the two occurrences of each letter suffices.

An alternative symbolic representation, sometimes encountered, gives the index -1 to a side whose direction conflicts with the reference direction. This representation gives, for example,

$$aa^{-1}bb^{-1}$$

in place of

$$+a-a+b-b.$$

Figure 11.4 is the rectangular plane diagram for a cylinder. Clearly, arbitrary directions may be assigned to sides x and y. All that is required is that x and y should appear alternately with the two occurrences of a, and that a should appear once with a positive sign and once with a negative sign. Thus, any cyclic permutation of

$$+a \pm x - a \pm y$$

suffices.

Similar representations may be obtained for the torus, the Moebius band, the Klein bottle, and the real projective plane by references to the corresponding plane diagrams. For each topologically distinct surface, there is a unique set of arrangements of signed letters, and thus equivalence classes of surfaces may be identified from symbolic representations. Further, certain particular sets of equivalence classes may be identified by specific types of arrangements of letters. For example, wherever each and every letter occurring appears exactly twice it is clear that the surface represented is closed. Further, noting that since sides are identified in pairs for closed surfaces, each such pair involves either like signs or unlike signs; if each and every pair consists of unlike signs the closed surface represented is orientable, otherwise it is non-orientable. The abstract study of sets of symbols together with re-arrangements which may be permitted without altering the 'value' of the overall expression (in this case, the type of manifold represented) is properly included in the branch of mathematics known as *combinatorics*.

12

The Standard Model

Removal of disc from a sphere—addition of handles—standard model of two-sided surfaces—addition of cross-caps—general standard model—rank—relation to Euler characteristic—decomposition of surfaces—general classification as open or closed, two-sided or one-sided—homeomorphic classes.

If a disc is removed from a sphere, it may be regarded as removed from a region of some map, as in Figure 12.1, or it may be regarded as a complete region of a map, as in Figure 12.2. In the case of Figure 12.1, removal of the disc may be regarded as equivalent to adding one complete arc to the map. In the case of Figure 12.2, it may be regarded as equivalent to reducing the number of regions of the map by one. In either event, the Euler characteristic is reduced by one. The sphere with a disc removed is homeomorphic to a disc.

Figure 12.3 depicts a sphere with two distinct discs removed. The removal of this second disc again results in the Euler characteristic being reduced by one. The sphere with two discs removed is homeomorphic to an open cylinder. This process of removing distinct discs from a

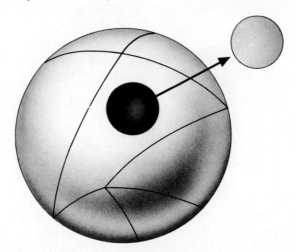

Fig. 12.1

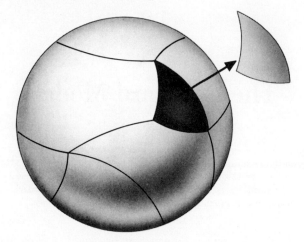

Fig. 12.2

sphere may be continued. Clearly, such a surface with r discs removed
has Euler characteristic

$$\chi = 2 - r;$$

it is termed a *sphere with r holes.*

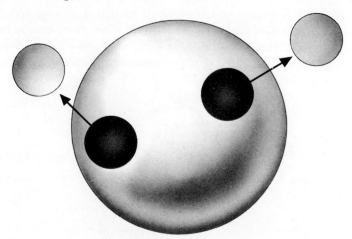

Fig. 12.3

Figure 12.4 depicts a cylinder, suitably deformed, attached to a
sphere with two holes so that its two end boundaries are exactly married
to the boundaries of the two holes. Attaching such a cylinder in this way
is termed adding a *handle*. A sphere with two holes to which a handle has

been added is homeomorphic to a torus. The process of adding handles to pairs of holes may be continued. Any number of handles may be added to a sphere with r holes provided that for p handles there are r holes available, where $r \geq 2p$. A sphere with two handles and no remaining holes is homeomorphic to a two-fold torus. Generally, a sphere with p handles and no remaining holes is homeomorphic to a p-fold torus.

Fig. 12.4

When a map is drawn on a sphere with p handles and no remaining holes, the surface may be topologically deformed so that no vertex and no arc segment of the map lies along any of the former boundaries where a handle has been rejoined on to the surface. Because of the requirement for every region of a map to be simply connected, it follows that at least one arc must travel lengthwise along every handle in the manner depicted in Figure 12.5.

Fig. 12.5

It is now intended to detach each handle at one end only, resulting in a sphere with p holes and p 'tubes' protruding from it. Such 'tubes' are termed *cuffs*. Before this is done, however, further vertices are added to the map, one at every intersection of an arc of the map and a boundary where a handle is to be detached. Any such boundary now becomes regarded as an arc of the map, as depicted in Figure 12.6. If the

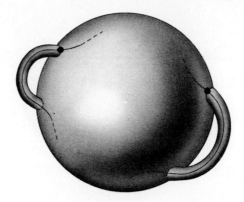

Fig. 12.6

number of additional vertices added to the map in the way just described is v, then the increase in the total number of arcs is $2v$ and the increase in the total number of regions is v. Thus, if the original map had V vertices, E arcs, and F regions, the modified map has $V+v$ vertices, $E+2v$ arcs, and $F+v$ regions. When the handles are detached at one end so as to form cuffs, the arcs and vertices on the boundaries are thereby duplicated. Thus, a further v vertices and v arcs are added, making a total of $V+2v$ vertices, $E+3v$ arcs, and $F+v$ regions.

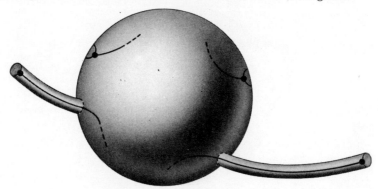

Fig. 12.7

The resulting surface may be made closed if the holes left by the removal of handle ends and the open cuff ends are all filled in by the addition of discs. Clearly, $2p$ discs are required to complete this task. The resulting surfaces is homeomorphic to a sphere, and has a map on it having $V+2v$ vertices, $E+3v$ arcs, and $F+v+2p$ regions, part of such a map being depicted in Figure 12.7.

Since the Euler characteristic of a sphere is equal to two, it follows that

$$(V+2v)-(E+3v)+(F+v+2p) = 2,$$

whence

$$V-E+F = 2-2p.$$

Thus, the Euler characteristic of a surface homeomorphic to a sphere with p handles is given by

$$\chi = 2-2p.$$

For a sphere with handles and holes, the Euler characteristic is given by

$$\chi = 2-2p-r.$$

Now the Euler characteristic of a disc is one, and that of a cylinder is zero. It would thus appear that the Euler characteristic of a sphere with p handles and r holes can be obtained by the addition or subtraction of the various surfaces which are respectively attached or removed. For example, a sphere with four handles and three holes has Euler characteristic

$$\chi = 2-2p-r$$
$$= 2-8-3 = -9,$$

and this can also be obtained as

$$\chi = \chi(\text{sphere}) - (2p+r)\,\chi(\text{disc}) + p\chi(\text{cylinder})$$
$$= 2-(11\times1)+(4\times0) = -9.$$

Further, a sphere can be reconstituted from a sphere with r holes together with r discs. It can also be reconstituted from p open cylinders. joined end to end so as to form one single cylinder, together with two discs. Addition of the corresponding Euler characteristics in each case yields respectively

$$\chi(\text{sphere}) = \chi(\text{sphere with } r \text{ holes}) + r\chi(\text{disc})$$
$$= (2-r)+r = 2$$

and

$$\chi(\text{sphere}) = p\chi(\text{cylinder}) + 2\chi(\text{disc})$$
$$= (p\times0)+(2\times1) = 2.$$

Generally, if n open surfaces $S_1, \ldots, S_n$ are joined together along boundaries so as to give some resulting surface, the Euler characteristic of this resulting surface is given by

$$\chi(\text{resulting surface}) = \chi(S_1) + \ldots + \chi(S_n).$$

All the surfaces obtained by removing discs from a sphere and adding handles are two-sided. Figure 12.8 repeats the plane diagram of the real projective plane discussed in Chapter 11. This rectangle may be topologically deformed into a sphere with a hole having a boundary corresponding to the original sides of the rectangle. Such a deformation is depicted in Figure 12.9. When the corresponding sides of the hole are

Fig. 12.8

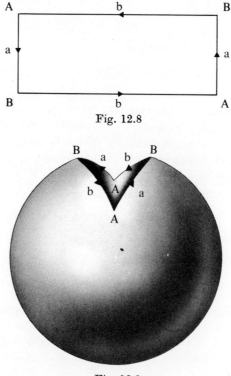

Fig. 12.9

married up, the directional senses being strictly observed, the result is a closed surface 'intersecting itself' in a line segment as depicted in Figure 12.10.

The closed surface of Figure 12.10 is one-sided. The end points of the line AB are single points, but all other points on AB are double

B

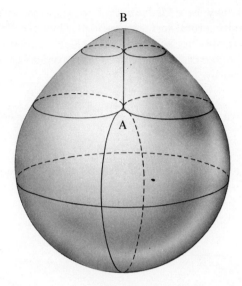

Fig. 12.10

points. If the lower hemisphere, which is homeomorphic to a disc, is now removed, a surface results which is termed a *cross-cap*. This is depicted in Figure 12.11. Since it is equivalent to the real projective plane with a disc removed, a cross-cap has Euler characteristic

$$\chi = \chi(\text{real projective plane}) - \chi(\text{disc})$$
$$= 1 - 1 = 0.$$

This value of χ is the same as that for a Moebius band. In fact, a cross-cap results if a Moebius band is deformed so that its boundary becomes a circle.

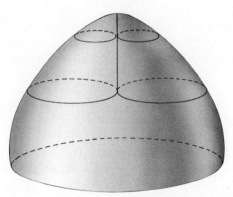

Fig. 12.11

Cross-caps may be added directly to a sphere with holes, thus making the resulting total surface one-sided. Since the Euler characteristic of a cross-cap is zero, and since one hole is required for each cross-cap, the Euler characteristic of a sphere with q cross-caps and r remaining holes is given by

$$\chi = 2 - q - r.$$

Overall, a sphere with p handles, q cross-caps, and r holes has Euler characteristic

$$\chi = 2 - 2p - q - r,$$

$2p + q + r$ being the total number of holes before the addition of the handles and cross-caps. A sphere with a handle, a cross-cap, and a hole is depicted in Figure 12.12.

Fig. 12.12

When a surface is described as a sphere with p handles, q cross-caps, and r holes, it is said to be presented in *standard model form*. A number of examples are given in the table following,

surface	p	q	r	χ
sphere	0	0	0	2
disc	0	0	1	1
real projective plane	0	1	0	1
Moebius band	0	1	1	0
cylinder	0	0	2	0
torus	1	0	0	0
two-fold torus	2	0	0	−2

In Chapter 5, the rank of an open surface was defined as the least number of cuts required to make the surface homeomorphic to a disc, and the rank of a closed surface was defined as that of the corresponding open surface obtained by the removal of a disc. It follows, therefore, that the rank of a surface whose standard model is a sphere with p handles, q cross-caps, and no holes is that of the open surface with a corresponding standard model having p handles, q cross-caps, and one hole.

The simplest example is provided by a sphere. Its standard model has

$$p = q = r = 0,$$

and its rank is that of a disc, for which

$$p = q = 0, r = 1.$$

The rank in each case is zero.

For a torus with a hole,

$$p = r = 1, q = 0.$$

To make this surface equivalent to a disc, two cuts are required. Such cuts are depicted in Figure 12.13. Once these cuts have been made, the

Fig. 12.13

surface may be deformed to give the pentagonal plane diagram depicted in Figure 12.14. This has the symbolic representation

$$+a+b-a-b \pm x,$$

the direction assigned to the side x being arbitrary. Since two cuts were

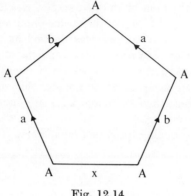

Fig. 12.14

required to reduce the surface to a disc, the rank of the torus with a hole is two, and the Euler characteristic is

$$\chi = 2-2p-q-r,$$
$$= 2-2-0-1 = -1.$$

This value is confirmed by the plane diagram of Figure 12.14 which has one distinct vertex, three distinct sides, and one face, giving

$$\chi = V-E+F$$
$$= 1-3+1 = -1.$$

Figure 12.15 depicts a sphere with one cross-cap and one hole. This may be cut once, as shown, and deformed to give the triangular

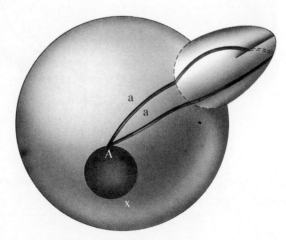

Fig. 12.15

plane diagram depicted in Figure 12.16. The symbolic representation of this surface is

$$+a+a \pm x.$$

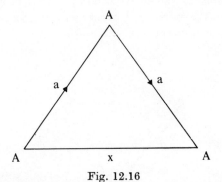

Fig. 12.16

The rank of this surface is one, and the Euler characteristic is

$$\chi = 2-2p-q-r$$
$$= 2-0-1-1 = 0,$$

or, alternatively,

$$\chi = V-E+F$$
$$= 1-2-1 = 0.$$

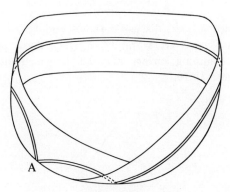

Fig. 12.17

The triangular plane diagram of Figure 12.16 may also be obtained by a single cut in a Moebius band as depicted in Figure 12.17. This confirms the values of p, q and r for a Moebius band already given in the standard model table.

Figure 12.18 depicts a sphere with one handle, one cross-cap, and one hole. This surface may be cut twice, as depicted by cuts d and e, so that the two 'protrusions' become detached. The handle, together with the portion of the sphere attached to it, may be deformed into a torus with a hole. The cross-cap, together with the portion of the sphere attached to it, is equivalent to a sphere with

$$q = r = 1, p = 0.$$

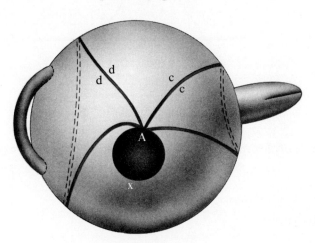

Fig. 12.18

These are the two surfaces discussed above, and whose plane diagrams are depicted in Figures 12.14 and 12.16. Once these 'protrusions' are removed, the remaining surface may be deformed into the triangle depicted in Figure 12.19.

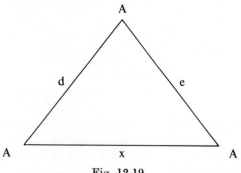

Fig. 12.19

Thus, the sphere with one handle, one cross-cap, and one hole may be decomposed into the three plane diagrams, depicted in Figure 12.20, whose sides correspond to the two cuts d and e, the boundary x of the hole, and cuts similar to those depicted in Figures 9.13 and 9.15. These plane diagrams have symbolic representations

$$+a+b-a-b-d,$$

$$+c+c-e,$$

$$+d+e \pm x.$$

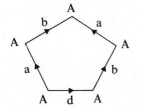

Fig. 12.20

The plane diagrams of Figure 12.20 may now be joined together by the identification of sides d and e, as in Figure 12.21. Once these sides have been married up and eliminated as boundaries, the resulting figure may be deformed in the plane to give the polygonal plane diagram of Figure 12.22, which has the symbolic representation

$$+a+b-a-b+c+c \pm x.$$

This diagram thus corresponds to the sphere with one handle, one cross-cap, and one hole, and the original surface may be reconstituted by suitable deformation of the polygonal plane diagram such that sides which are identified are married up in conformity with the arrow-head

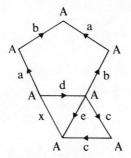

Fig. 12.21

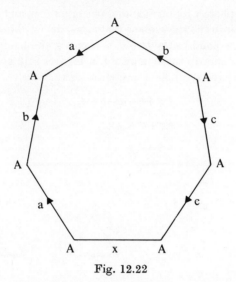

Fig. 12.22

directions. (It is not, however, possible to carry out this deformation in its entirety in ordinary three-dimensional space.) The polygon of Figure 12.22 has one distinct vertex, four distinct sides, and one face, giving

$$\chi = 1-4+1 = -2.$$

This agrees with the value of χ obtained from the numbers of handles, cross-caps and holes.

A two-way process has now been established. Polygonal plane diagrams may be obtained from standard models, and surfaces may be built up from polygonal plane diagrams. When a surface is built up in this way, it is clear that the addition of every plane diagram equivalent to a handle increases the rank by two, and the addition of every plane diagram equivalent to either a cross-cap or a hole increases the rank by one. Thus, for a closed surface, the rank is

$$2p+q = 2-\chi,$$

and for an open surface,

$$2p+q+r-1 = 1-\chi.$$

If a closed surface is also two-sided, its rank is $2p$ and its Euler characteristic is $2-2p$. From Chapter 6, p. 52

$$\chi = 2-2g,$$

where g is the genus. Hence, for a closed two-sided surface, the genus is equal to the number of handles of its standard model.

Surfaces may be classified by reference to the symbolic representations of their corresponding polygonal plane diagrams. Thus, if a symbolic representation contains at least one term which occurs only once, the surface represented must be open. For example, a symbolic representation of a cross-cap is

$$+a+a \pm x.$$

This is an open surface, and hence at least one symbol, in this case just x, appears once only. If, on the other hand, a symbolic expression is made up entirely of terms which occur twice, the surface represented must be closed. For example, a symbolic representation of a torus is

$$+a+b-a-b,$$

and here each of the two terms a and b appear twice. The signs of the terms are not of relevance in determining whether a surface is open or closed.

In the case of closed surfaces, if each term of a symbolic representation occurs once with positive sign and once with negative sign, the surface represented must be two-sided, that is, its standard model does not include a cross-cap. Similar considerations apply to the terms which appear twice in the symbolic representations of open surfaces. Thus, the representation

$$+a+b-a-b \pm x$$

denotes an open two-sided surface, whilst the representation

$$+a+b-a-b+c+d-c-d$$

represents a closed two-sided surface. Since the standard model of any one-sided surface must include a cross-cap, the symbolic representation of such a surface must include a term appearing twice, on each occasion with the same sign. The representations

$$+a+b-a+b$$

and

$$+a+b-a-b+c+c \pm x,$$

for example, both represent one-sided surfaces, the former being closed and the latter open.

The classification procedure described above leads to four distinct classes of surface, namely:

closed two-sided surfaces, for which $p \geqq 0$, $q = r = 0$;

closed one-sided surfaces, for which $q > 0$, $r = 0$;

open one-sided surfaces; for which $q > 0$, $r > 0$;

open two-sided surfaces, for which $q = 0$, $r > 0$.

If two surfaces are homeomorphic they must clearly belong to the same class from the four listed above. In addition, their corresponding values for p, q and r must be respectively the same, and their symbolic representations must therefore be identified according to certain prescribed rules. This topic admits of very considerable general development under a special topological study termed *combinatorial topology*.

13

Continuity

Preservation of neighbourhood—distance—continuous and discontinuous curves—formal definition of distance—triangle inequality—distance in n-dimensional Euclidean space—formal definition of neighbourhood—ε-δ definition of continuity at a point—definition of continuous transformation.

In Chapter 3 the concept of a *one–one bicontinuous transformation* was introduced, and this concept has underpinned the whole idea of topological equivalence. Thus, two figures are homeomorphic if one can be transformed into the other by any one–one bicontinuous transformation. The intuitive approach to such transformations is clearly inadequate for any precise development of topological concepts, though it is useful where it is intended merely to give a general introduction to the kinds of topics which are studied under the heading of 'topology'. By requiring that such transformations and their inverses preserve neighbourhoods, in that points which are in some sense 'near' remain 'near', only a very imprecise definition is in fact presented, because the word 'near' needs to be precisely defined in mathematical language. Clearly, in order to define 'near', some sort of understanding of 'distance' is also required, and, whilst this is easily understood in the familiar context of three-dimensional Euclidean space, it is not so obvious when the more abstract spaces of the mathematician are being considered. It is also true that the neighbourhood approach, even when precisely defined in terms of distance, is too restrictive for the needs of the topologist, since it is often required to consider sets for which any intuitive understanding of distance, and hence of nearness, would be largely meaningless. To make a first intuitive approach to *continuity* is, however, generally helpful.

Figure 13.1 depicts a *continuous* curve in a plane. If axes are added, as in Figure 13.2, then the curve defines a one–one bicontinuous transformation of the set of real numbers to itself. The transformation is *one–one* because, given any point x, a single point y is defined as the image of that x and of no other x, in the way indicated. The transformation is *bicontinuous*, because any point 'near' x is mapped to a point 'near' y by the transformation, and any point 'near' x by its

Fig. 13.1

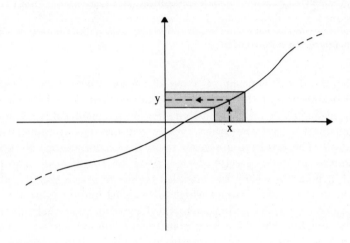

Fig. 13.2

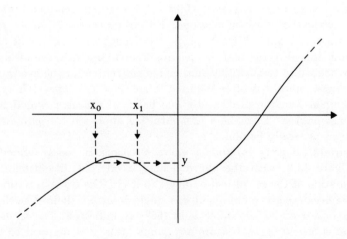

Fig. 13.3

inverse. The x- and y-axes both represent the set of real numbers **R**. A number 'near' x may intuitively be regarded as lying within some prescribed distance from x on the axis, as depicted in Figure 13.2.

Figure 13.3 depicts a transformation that is not one–one, although the original curve is very similar to that of Figures 13.1 and 13.2. The difference between these curves is, however, very important. The curve of Figures 13.1 and 13.2 increases throughout, that is, any increase in x necessarily involves an increase in y. However, that of Figure 13.3 does not so increase. It reaches a peak, then temporarily decreases before finally increasing again. As a consequence, it is possible to find pairs of points, for example the pair (x_0, x_1), each mapping to a single point y. Such a transformation is termed *many–one*. The curve is again continuous, however, and a point 'near' to any point x is still mapped to a point 'near' to the point y corresponding to x.

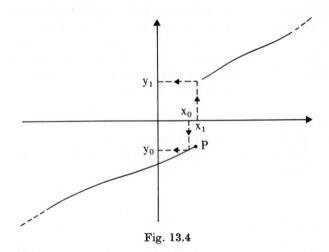

Fig. 13.4

Figure 13.4 depicts what can intuitively be seen to be a *discontinuous* curve. This particular curve increases everywhere so that it still represents a one–one transformation, but the neighbourhood preserving property of the earlier transformations is now lost. The points x_0 and x_1 may be thought of as in some sense 'near' each other, but the two points y_0, y_1 to which they are respectively mapped are clearly not 'near' in the same sense. In the first place, a continuous progress from the value x_0 to the value x_1 involves a sudden jump in the corresponding progress from y_0 to y_1. This jump takes place immediately *after* the point P indicated on the curve, that is, there is a continuous increase in y as x increases up to and including the values corresponding to P, thereafter as x further increases continuously there is the jump in the value of y. In the second place, for the inverse transformation, there are values of y for which no corresponding values of x are defined, namely those values of y which are in the 'gap' where the jump in the curve occurs.

In order to be mathematically precise about 'nearness' and ultimately about 'continuity' also, it is necessary first to be precise about 'distance'. If x_0, x_1 are two points on an axis representing the set of real numbers $\mathbf{R}$, the *distance* between x_0 and x_1 is defined as

$$d(x_0, x_1) = |x_0 - x_1|,$$

that is, as the modulus of the difference between the two real numbers being represented. It is now possible to specify 'nearness' by requiring that for two points considered 'near' each other the distance between them must be less than some prescribed positive quantity.

It is clear that distance as defined above satisfies certain conditions. In the first place, the distance between any two distinct points must be a positive quantity. Secondly, if the distance between two points is zero then it follows that the points are not distinct, but are one and the same point. Thirdly, the distance from any point x_0 to another point x_1 is by definition the same as the distance from x_1 to x_0; the direction of travel is irrelevant. Fourthly, distances may be added so that the sum of the distances from a point x_0 to a point x_1 and from a point x_1 to a point x_2 gives the total distance from x_0 to x_2 provided that the same direction of travel is maintained, that is, provided that x_1 lies between x_0 and x_2. This is depicted in Figure 13.5.

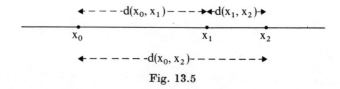

Fig. 13.5

It is not necessary, however, to confine the concept of distance to the one dimensional case of points lying on an axis representing the set of real numbers $\mathbf{R}$. If P_0, P_1 are two points lying in a plane with co-ordinates (x_0, y_0), (x_1, y_1) relative to a set of rectangular Cartesian axes, then the distance from P_0 to P_1 is defined as

$$d(P_0, P_1) = \sqrt{[(x_0 - x_1)^2 + (y_0 - y_1)^2]},$$

where the positive square root is to be assumed. This formula is an expression for Pythagoras' theorem, as may be seen from Figure 13.6.

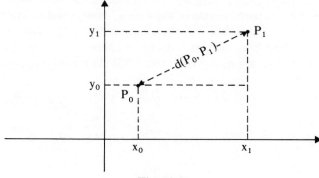

Fig. 13.6

Distance, as now defined, satisfies the same first three conditions of the one-dimensional case, but, instead of being simply additive, it now satisfies the *triangle inequality*

$$d(P_0, P_1) + d(P_1, P_2) \geq d(P_0, P_2).$$

This is an expression of the familiar geometric theorem which states that the sum of the lengths of two sides of a triangle is greater than the length of the third side, with the equality allowing for the degenerate case where all three points lie on a straight line. The one-dimensional case also satisfies this inequality. The triangle inequality is depicted in Figure 13.7. It will be seen that because the expression is an inequality, there is no longer any requirement for a restriction on the relative positions of the three points P_0, P_1, P_2 in the plane.

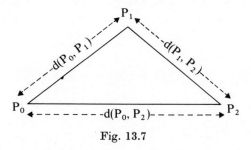

Fig. 13.7

The two-dimensional expression for distance can be generalized to three or more dimensions. Thus, in the three dimensional case, distance is defined by

$$d(P_0, P_1) = \sqrt{[(x_0 - x_1)^2 + (y_0 - y_1)^2 + (z_0 - z_1)^2]},$$

where (x_0, y_0, z_0), (x_1, y_1, z_1) are the respective co-ordinates of P_0, P_1 relative to an appropriate set of three axes, as depicted in Figure 13.8.

In each of the definitions of distance given so far, the context in which the definition has been made is an n-dimensional Euclidean space, in which each axis represents the set **R**. This is a far too restrictive

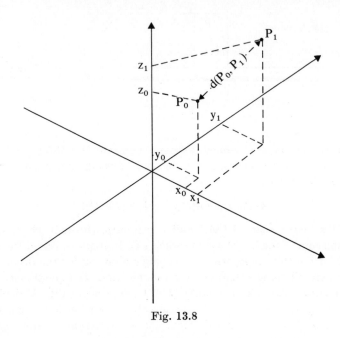

Fig. 13.8

context for the purposes of topology, though it provides a useful starting point. The emphasis, however, should be placed, not on the formula for calculating distances, but on the properties satisfied, namely:

$$d(P_0, P_1) \geqq 0;$$

$$d(P_0, P_1) = 0 \text{ if and only if } P_0 = P_1;$$

$$d(P_0, P_1) = d(P_1, P_0);$$

$$d(P_0, P_1) + d(P_1, P_2) \geqq (dP_0, P_2).$$

Still remaining within the context of n-dimensional Euclidea space, it is now possible to give a more precise definition of 'nearness' and hence also of 'continuity'. If x_0, x_1 are points on a real axis (th is, an axis representing the set **R**) then x_0 and x_1 can be said to δ-near if

$$d(x_0, x_1) < \delta,$$

where δ is some prescribed positive real number. The set of all points x_1 for which the given condition is satisfied is termed a *neighbourhood* (or *open ball*) of x_0.

The intuitive definition of 'continuity' can now be made more precise. Given a many–one or one–one transformation which maps the set **R** to itself (or a subset of itself), such as that depicted in Figure 13.9, the transformation is said to be *continuous at the point* x_0 if, given any positive quantity ε, however small, there exists a positive quantity δ such that for all points x_1 for which

$$d(x_0, x_1) < \delta,$$

it is true that

$$d(y_0, y_1) < \varepsilon,$$

where y_0, y_1 are the points to which x_0, x_1 respectively are mapped under the given transformation. Thus, the neighbourhood of x_0 is mapped to a neighbourhood of y_0.

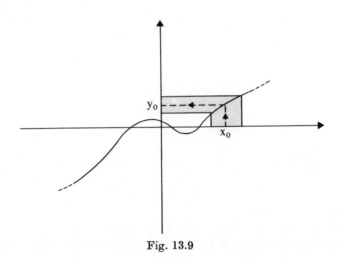

Fig. 13.9

Figure 13.10 repeats the discontinuous transformation of Figure 13.4. If ε is first chosen as depicted, then it is not possible to find any positive quantity δ satisfying the prescribed condition for continuity. The condition is satisfied for points x_1 lying to the left of x_0 on the axis, but if x_1 lies to the right of x_0, no matter how close, the point y_1 to which it is mapped is necessarily at a distance greater than ε from y_0 at x_0. It is, however, continuous everywhere else. If, for example, the point x_0' is considered, where $x_0' < x_0$, then, however small the value of ε is chosen to be, there is always some δ such that for all points x, for which

$$d(x_0', x_1) < \delta, \text{ then } d(y_0', y_1) < \varepsilon.$$

It is important to note that it is the quantity ε which is first specified, and that the condition for continuity at a point is then defined in terms of the existence of the quantity δ.

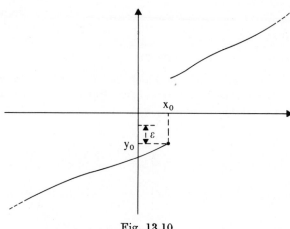

Fig. 13.10

A transformation is said to be *continuous* if it is continuous at every point x of its domain, where the *domain* of a transformation is the set of all points which are to be mapped by the transformation concerned. A transformation is said to be bicontinuous if it is continuous 'both ways', that is, if it is itself continuous and if its inverse is a continuous transformation also.

So far, the language of set theory has been kept to a minimum. It now becomes necessary however to extend the use of this language to a certain extent in order to present topological concepts precisely and in a form which is not restricted to the context of n-dimensional Euclidean space. The next chapter is therefore devoted to a brief resumé of those concepts and terms from set theory which will be required in subsequent chapters as some of the concepts of topology are presented in formal mathematical language. Readers who are familiar with set-theoretical language may proceed directly to Chapter 15.

14

The Language of Sets

Sets and subsets defined—set equality—null set—power set—union and intersection—complement—laws of set theory—Venn diagrams —index set—infinite sets—intervals—Cartesian product—n-dimensional Euclidean space.

In a number of the preceding chapters, the words 'set' and 'subset' have been used in contexts where their meanings should have been intuitively clear. For example, in the proof of the Jordan curve theorem for a polygonal path (see Chapter 9), the set of points of the plane not belonging to the path itself was divided into two disjoint subsets, subsequently defined as the subsets of points outside and inside the path respectively. In this instance, the term 'set' was used to denote the collection of all points in the plane, and the term 'subset' to denote a collection consisting of some but not all of the points belonging to the original set.

A *set* is simply a collection of distinct objects such that, given any object whatever, it is possible to determine whether or not the given object belongs to that set. Thus, a typical set is the set of all integers **Z**. Given any object whatever, it is possible to determine whether or not it belongs to the set **Z**. For example, given the objects:

lemon, moon, 21, 2·75, 3/2,

it is possible to see that only the number 21 belongs to **Z**, the remaining four objects not belonging to **Z**.

The expression

$$x \in X$$

is the symbolic representation of the statement 'the element x belongs to the set X', whilst the expression

$$x \notin X$$

is the symbolic representation of the statement 'the elements x does not belong to the set X'. The elements belonging to a set may be listed or described. Thus, alternative representations of the set of integers from 1 to 9 are

$$\{1, 2, 3, 4, 5, 6, 7, 8, 9\},$$

often abbreviated to

$$\{1, 2, \ldots, 9\},$$

and

$$\{x \colon x \text{ is an integer and } 1 \leq x \leq 9\},$$

read as 'the set of elements x, such that x is an integer and one is less than or equal to x, which is less than or equal to nine'. Two sets are equal if and only if they comprise exactly the same elements.

If every element of the set X is also an element of the set Y, then the set X is a *subset* of the set Y. This is written

$$X \subseteq Y,$$

in which case Y is also a *superset* of X. If, in addition, X and Y are not equal, then X is a *proper subset* of Y, written

$$X \subset Y.$$

Thus, every set is a subset but not a proper subset of itself. Clearly, also,

$$\text{if } X \subseteq Y \text{ and } Y \subseteq X, \text{ then } X = Y.$$

The *null set* (or the *empty set*) is the set having no elements, and is denoted by the Scandinavian letter $\varnothing$. The null set is a subset of every set, including itself; that is $\varnothing \subseteq X$ for every set X. It is usually regarded as an improper subset of any set X.

The *power set* of a set X is the set of all subsets of X. Thus, if

$$X = \{x_0, x_1, x_2\},$$

then the power set of X is given by

$$\mathscr{P}(X) = \{X, \{x_0, x_1\}, \{x_0, x_2\}, \{x_1, x_2\}, \{x_0\}, \{x_1\}, \{x_2\}, \varnothing\}.$$

Clearly, if X has a finite number of elements, n, say, then the power set of X will consist of 2^n subsets of X.

The *union* of two sets, X and Y, is the set whose elements are either elements of X, or of Y, or of both X and Y. Thus, if

$$X = \{a, b, c, d, e\}, \ Y = \{d, e, f, g\},$$

the union of X and Y, written $X \cup Y$, is given by

$$X \cup Y = \{a, b, c, d, e, f, g\}.$$

The *intersection* of two sets, X and Y, is the set whose elements are elements both of X and of Y. Thus for the two sets X and Y defined above, the intersection of X and Y, written $X \cap Y$, is given by

$$X \cap Y = \{d, e\}.$$

If there are no elements common to both X and Y, then $X \cap Y = \varnothing$, and they are said to be *disjoint*.

The union and intersection of sets may often be usefully illustrated by means of *Venn diagrams*. Figure 14.1 is a Venn diagram in which sets X and Y are depicted by intersecting circles, and the shaded area

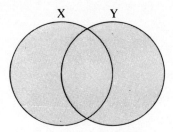

Fig. 14.1

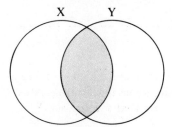

Fig. 14.2

represents $X \cup Y$. Similarly, the shaded area of Figure 14.2 represents $X \cap Y$. Figure 14.3 depicts two sets which are disjoint. In all these Venn diagrams, the set of elements belonging to X and Y is represented by the areas inside the respective circles labelled X and Y.

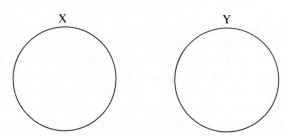

Fig. 14.3

The *complement* of a set X, written X', is the set of all elements not belonging to X. Usually, the total set of elements being considered is restricted to some overall set, termed the *universal set* U, and the complement of any set X is defined relative to this universal set U.

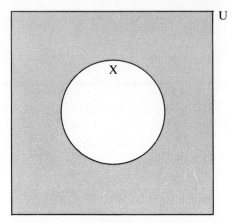

Fig. 14.4

The shaded area of the Venn diagram of Figure 14.4 represents the complement of X with respect to the universal set U. From the definition of complement it follows that for any set X,

$$X \cup X' = U$$

and that

$$X \cap X' = \varnothing.$$

The *relative complement* of a set X in a set Y is the set of all elements of Y which are not also elements of X. The relative complement of X in Y is often written using the difference sign as

$$Y - X.$$

Clearly,

$$Y - X = Y \cap X'.$$

An illustration of set union, intersection and complementation is provided by a continuous closed curve on a plane surface. The universal set U of all points of the plane may be divided into three disjoint subsets, the set C of points belonging to the curve, the set X of points inside the curve, and the set Y of points outside the curve. This is depicted in Figure 14.5. The set of points of the plane which are not inside the curve is the complement of X, given by

$$X' = U - X$$

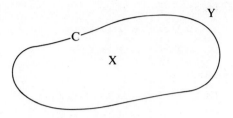

Fig. 14.5

and the set of points of the plane which are not outside the curve is the complement of Y, given by

$$Y' = U - Y.$$

The sets X' and Y' are not respectively equal to Y and X. Y and X are proper subsets of X' and Y' respectively; thus,

$$Y \subset X' \text{ and } X \subset Y'.$$

Further, despite the fact that X and Y are disjoint, that is,

$$X \cap Y = \varnothing,$$

X' and Y' are not disjoint since both include the set C of points belonging to the curve, thus,

$$X' \cap Y' = C.$$

In a similar manner, the regions of a map on a surface are disjoint only if their boundaries are excluded. Thus, if regions F_1, F_2 have a common boundary, included in both F_1 and F_2, that common boundary is given by the intersection

$$F_1 \cap F_2.$$

The union of all such pairwise intersections, which may be written

$$\bigcup (F_i \cap F_j)$$

is the set of all points of the network comprising the boundaries of the regions only.

The set operations of union, intersection and complementation satisfy a great number of laws, the most fundamental of which are given below:

Idempotent laws:

$$X \cup X = X,$$
$$X \cap X = X;$$

Commutative laws:

$$X \cup Y = Y \cup X,$$
$$X \cap Y = Y \cap X;$$

Associative laws:

$$(X \cup Y) \cup Z = X \cup (Y \cup Z),$$
$$(X \cap Y) \cap Z = X \cap (Y \cap Z);$$

Distributive laws:

$$X \cup (Y \cap Z) = (X \cup Y) \cap (X \cup Z),$$
$$X \cap (Y \cup Z) = (X \cap Y) \cup (X \cap Z);$$

Identity laws:

$$X \cup \varnothing = X,$$
$$X \cap \varnothing = \varnothing,$$
$$X \cup U = U,$$
$$X \cap U = X;$$

Complementation laws:

$$X \cup X' = U,$$
$$X \cap X' = \varnothing,$$
$$(X')' = X,$$
$$\varnothing' = U,$$
$$U' = \varnothing;$$

De Morgan's laws:

$$(X \cup Y)' = X' \cap Y',$$
$$(X \cap Y)' = X' \cup Y'.$$

If X is a subset of Y, that is if $X \subseteq Y$, then the following expressions are all equivalent and true:

$$X \cup Y = Y,$$
$$X \cap Y = X,$$
$$Y' \subseteq X',$$
$$X' \cup Y = U,$$
$$X \cap Y' = \varnothing,$$
$$X' \cup Y' = X',$$
$$X' \cap Y' = Y'.$$

These relationships can more easily be seen by reference to the Venn diagram of Figure 14.6, where the set X' is denoted by vertical shading and the set Y' by horizontal shading.

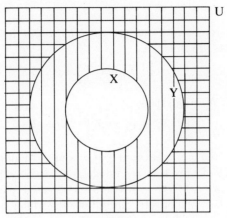

Fig. 14.6

The symbolic representation for the union of a number of sets, X_i, has already been introduced above, and may be written as

$$\bigcup_i X_i.$$

The corresponding symbolic representation for the intersection of a number of sets, X_i, is

$$\bigcap_i X_i.$$

In order to make these symbolic representations precise, it is necessary to introduce the concept of an *index set*, the elements of which may be considered simply as names or labels. Given a collection of sets (or subsets of a set), any one may be identified if to each is assigned a unique element of the index set. Thus, if a collections of eight sets is to be indexed, the index set may be the set of integers

$$\{0, 1, 2, 3, 4, 5, 6, 7\},$$

or the set of letters

$$\{a, b, c, d, e, f, g, h\},$$

or any other suitable set consisting of eight elements. The two just suggested would give, for example,

$$\{X_0, X_1, \ldots, X_7\}$$

or

$$\{X_a, X_b, \ldots, X_h\}.$$

The collection of sets to be indexed is thus put in one–one correspondence with the elements of the index set chosen. Where a collection
of sets is indexed by means of a set M, then this is represented symbolically by

$$\{X_i : i \in M\}.$$

The symbolic representations for the union and intersection of the
collection of sets are thus

$$\bigcup_{i \in M} X_i \text{ and } \bigcap_{i \in M} X_i$$

respectively. If $M = \varnothing$, then

$$\bigcup_{i \in M} X_i = \varnothing \text{ and } \bigcap_{i \in M} X_i = U.$$

The most frequently encountered sets used to index a collection of
sets are the set of non-negative integers, $\{0, 1, 2, \ldots\}$, and the set of
positive integers, $\{1, 2, 3, \ldots, \}$. These may be used to index infinite sets
and infinite collections of sets. An infinite set which can be so indexed
is said to be *denumerable* (or *countable*). The set of all rational numbers is
an example of a denumerable set, though it is not immediately obvious
that it can be put into one–one correspondence with the positive integers.
The set of all real numbers is not denumerable, neither is any interval of
the set of real numbers, such as the *unit interval* $[0, 1]$.

The *Cartesian product* of two sets, X and Y, is the set of all *ordered
pairs* of elements (x, y), such that x is an element of X, and y is an
element of Y. For example, if

$$X = \{a, b, c, d\}, \ Y = \{0, 1\},$$

then the Cartesian product of X and Y, denoted by $X \times Y$, is given by

$$X \times Y = \{(a, 0), (a, 1), (b, 0), (b, 1), (c, 0), (c, 1), (d, 0), (d, 1)\}.$$

Clearly, if X has m elements and Y has n elements, then $X \times Y$ will
have mn elements. For the sets X, Y above, the Cartesian product
$Y \times X$ is given by

$$Y \times X = \{(0, a), (0, b), (0, c), (0, d), (1, a), (1, b), (1, c), (1, d)\}.$$

This is not the same set as $X \times Y$ because the ordering of the elements
within the pairs is taken into consideration. If $X \times Y = Y \times X$ then it
follows that $X = Y$.

In the same way the Cartesian product of three sets, X, Y, Z, is
the set of all ordered triples (x, y, z), such that $x \in X$, $y \in Y$, $z \in Z$. This
concept may be generalized to an infinite number of sets by using the
set of positive integers $\mathbf{Z}$ as an index set, and writing

$$\prod_{i \in Z} X_i.$$

If X and Y are each the set of all real numbers $\mathbf{R}$, then the Cartesian plane will be represented by the Cartesian product $\mathbf{R} \times \mathbf{R}$. This will be denoted by $\mathbf{R}^2$. Generalizing this, gives

$$\mathbf{R}^n = \underbrace{\mathbf{R} \times \ldots \times \mathbf{R}}_{n \text{ times}}$$

as the set of all points in n-dimensional Euclidean space, each point being identified by an n-tuple of real numbers $(x_1, \ldots, x_n)$. It is important to remember, however, that the formal definition of n-dimensional Euclidean space also includes the definition of the generalized distance function (see page 122).

15

Functions

Definition of function—domain and codomain—image and image set—injection, bijection, surjection—examples of functions as transformations—complex functions—inversion—point at infinity—bilinear functions—inverse functions—identity function—open, closed, and half-open subsets of **R**—tearing by discontinuous functions.

A *function* is a rule or correspondence which assigns to each element of a given set, known as the *domain* of the function, an element of a second set, known as the *codomain* of the function. Thus a function

$$f: X \to Y$$

assigns to each $x \in X$ a single element $y \in Y$. In the example shown in Figure 15.1, the domain of the function is the set $\{a, b, c, d, e\}$ and the codomain is the set $\{1, 2, 3\}$.

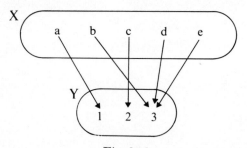

Fig. 15.1

The element of Y which corresponds to a given element x of X is termed the *value* of the function at x, and also the *image* of the element x under the function f. It is usually denoted by $f(x)$. Thus, for the example of Figure 15.1,

$$f(a) = 1, f(b) = 3, f(c) = 2, f(d) = 3, f(e) = 3.$$

The set of all values $f(x)$, $x \in X$, is called the *image set* and is denoted by $f(X)$. Thus, in this instance,

$$f(X) = \{1, 2, 3\} = Y.$$

The image set must necessarily be a subset of the codomain of a function; it does not, however, have to be equal to the codomain. Figure 15.2 depicts a function whose image set is a proper subset of its codomain. It is important to note, however, that although there may be elements

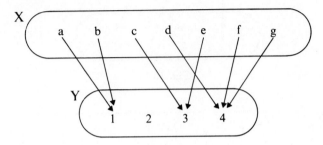

Fig. 15.2

of the codomain which are not the image of any element of the domain (as when the image set is a proper subset of the codomain), there must be no element of the domain without a corresponding image.

If a function $f: X \to Y$ assigns elements of Y to elements of X in such a way that no element of Y is the image of more that one element of X, then the function is termed an *injection*. Figure 15.3 depicts an

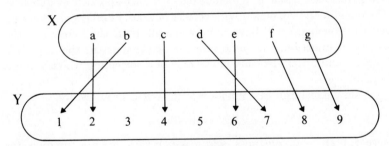

Fig. 15.3

injection $f: X \to Y$. The correspondence of Figure 15.3 is one–one, but the image set is not equal to the codomain of the function. If the correspondence is one–one and if at the same time $f(X) = Y$, then the function is termed a *bijection*. Figure 15.4 depicts a bijection $f: X \to Y$. If, on the other hand, $f(X) = Y$, but the correspondence is many–one, that is, some $y \in Y$ is the image of more than one $x \in X$, the function is called a *surjection*. Figure 15.5 depicts such a surjection, $f: X \to Y$ (as also does Figure 15.1). (Most mathematicians regard

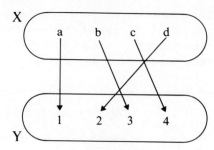

Fig. 15.4

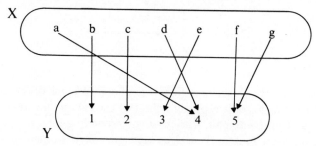

Fig. 15.5

bijections as a special case of surjections, i.e. the expression 'many-one correspondences' is taken to include one–one correspondences.) Clearly, if the domain and codomain both consist of finite numbers of elements, then the codomain will have more elements than the domain in the case of an injection, the same number of elements in the case of a bijection, and fewer elements in the case of a surjection which is not a bijection.

A simple example of a function $f: \mathbf{R} \to \mathbf{R}$ is provided by

$$f(x) = ax+b,$$

where a, b are real constants, and a is non-zero. Such a function, for example the one shown in Figure 15.6, is a bijection, since it is one–one, and the image set and the codomain are the same, that is $X = Y = \mathbf{R}$. If, however, the constant a is zero, then the function $f: X \to Y$ becomes many–one and is therefore no longer a bijection. In this case, every point x of the domain maps to the one point b of the codomain, and the image set becomes the proper subset of the codomain consisting of the one element b, as shown in Figure 15.7. In this case, the function is neither an injection, nor a bijection, nor a surjection. It can be made a surjection, however, by redefining the codomain from the start as the image set $\{b\}$.

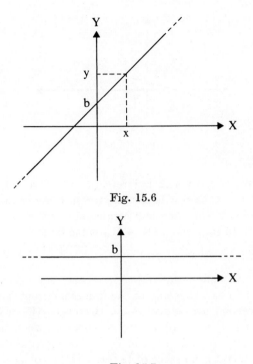

Fig. 15.6

Fig. 15.7

The particular bijection of Figure 15.6 provides an example of a linear transformation which stretches or compresses without tearing. Figure 15.8 shows how a bijection stretches a subset X of $\mathbf{R}$ onto its image $f(X)$. Figure 15.9 shows a similar situation, but in this case X is compressed rather than stretched. Clearly, for the bijection $f: \mathbf{R} \to \mathbf{R}$, given by $f(x) = ax + b$, it is the value of the constant a that determines whether a given interval X of $\mathbf{R}$ is stretched or compressed onto its

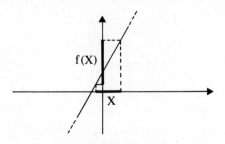

Fig. 15.8

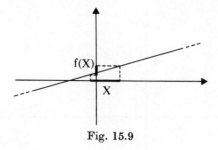

Fig. 15.9

image $f(X)$. If $a > 1$, then X is stretched, and if $a < 1$, then X is compressed. If $a = 1$, then X is neither stretched nor compressed.

An example of stretching and compressing which is not linear is shown in Figure 15.10, where $f : \mathbf{R} \to \mathbf{R}_0^+$ is the function

$$f(x) = x^2.$$

($\mathbf{R}_0^+$ is the set of all non-negative real numbers.) Since each element of $\mathbf{R}_0^+$ is assigned to two elements of the domain, except in the case of zero, the correspondence is many–one. However, $f(\mathbf{R}) = \mathbf{R}_0^+$, so the function is a surjection. It will readily be seen that whether or not any particular interval $X \subset \mathbf{R}$ is stretched or compressed depends on whether the interval is a subset of $\{x : |x| \geq 1\}$, or of $\{x : |x| \leq 1\}$. In Figure 15.10 the subset X_0 is compressed whilst the subset X_1 is stretched.

In considering whether or not a given interval is stretched or compressed it is necessary to consider respectively those intervals

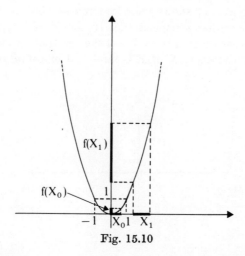

Fig. 15.10

lying wholly in $1 \leqq x < \infty$ or $-\infty < x \leqq -1$, and those lying in $0 \leqq x \leqq 1$ or $-1 \leqq x \leqq 0$. In other words, the interval must be selected so that it lies either wholly in the area of stretching or wholly in the area of compression, and also so that the transformation is effectively one–one. If, for example, an interval were considered which included both positive and negative values of x, then, although the condition for continuity may be satisfied, there would now be a many-one situation where pairs of elements of the domain would map to a single element of the codomain. In the context of topological transformations, such coalescence of points is not permitted.

The first three chapters considered the way in which the equivalence classes of the various geometries could be determined in terms of permitted transformations. Congruence classes, for example, were defined as equivalence classes obtained when only rigid transformations are permitted between elements belonging to the same class. By contrast, topological equivalence classes allow elastic deformations such as bending and stretching, or even cutting provided that the cut is subsequently exactly repaired.

The rigid transformation of translation in ordinary three-dimensional Euclidean space is a bijection, $f: \mathbf{R}^3 \to \mathbf{R}^3$, such that each point x of $\mathbf{R}^3$, given by the ordered triple (x_1, x_2, x_3), is mapped to a unique point $f(x)$ of $\mathbf{R}^3$, given by the ordered triple $(x_1+a_1, x_2+a_2, x_3+a_3)$, where a_1, a_2, a_3 are real constants. The rigid transformation of rotation in $\mathbf{R}^2$ is a bijection, $f: \mathbf{R}^2 \to \mathbf{R}^2$, such that one point of $\mathbf{R}$, termed the *centre* of rotation, is mapped to itself, whilst all other points of $\mathbf{R}^2$ are mapped so that each and every ray from the centre of rotation is mapped to a corresponding ray of a fixed angle from it. If the centre of rotation is taken as the origin of a polar co-ordinate system (r, θ), then f maps according to the rule,

$$(r, \theta) \mapsto (r, \theta + \alpha)$$

where the fixed angle α is termed the *angle of rotation*. The rigid transformation of reflection in $\mathbf{R}^3$ is a bijection, $f: \mathbf{R}^3 \to \mathbf{R}^3$ which maps each point on some line L in $\mathbf{R}^3$ to itself, whilst each point of $\mathbf{R}^3$ not belonging to L is mapped to the corresponding point perpendicularly opposite to it with respect to the line L.

When the equivalence classes are such that similar figures are within one and the same class, magnification and contraction are permitted transformations. The elastic transformation of magnification in $\mathbf{R}^3$ is a bijection, $f: \mathbf{R}^3 \to \mathbf{R}^3$, which maps one point of $\mathbf{R}^3$, termed the *centre of magnification*, to itself, whilst all other points are mapped so that the image of every point lies on the same ray as the point itself but at a distance greater (or smaller in the case of contraction) from the centre of magnification. If the centre of magnification is taken as the origin of a rectangular Cartesian co-ordinate system, then the bijection $f : \mathbf{R}^3 \to \mathbf{R}^3$ maps according to the rule

$$f : (x_1, x_2, x_3) \mapsto (ax_1, ax_2, ax_3),$$

where a is a positive real constant greater than one in the case of magnification, and less than one in the case of contraction. If the centre of magnification is taken as the origin of a conventional spherical co-ordinate system (r, θ, ϕ), then f maps according to the rule

$$(r, \theta, \phi) \mapsto (ar, \theta, \phi),$$

It has already been seen that where the domain and codomain of a function are both identifiable with the set of real numbers $\mathbf{R}$ (or with subsets of $\mathbf{R}$), it is possible to give a pictorial representation of the function as was done, for example, in Figures 15.6 to 15.10. Such simple pictorial representation is, however, confined to functions of a single real variable.

Functions of a complex variable are functions where the domain and codomain are both identifiable with the set of complex numbers $\mathbf{C}$, which is, in turn identifiable with $\mathbf{R}^2$. Thus, the complex number $z = x + iy$, where $i^2 = -1$, is identifiable with the ordered pair of real numbers (x, y). It is usual to express the rule for a function $f : \mathbf{C} \to \mathbf{C}$ in the form

$$w = f(z),$$

where

$$w = u + iv, \text{ and } z = x + iy.$$

Any particular subset of $\mathbf{C}$ taken as the two-dimensional equivalent of an interval of $\mathbf{R}$, then becomes a subset of a plane representing $\mathbf{R}^2$ with x- and y-axes, and the corresponding image set of the codomain of a function f is then a subset of a second plane, also representing $\mathbf{R}^2$ but with u- and v-axes.

As an example, Figure 15.11 depicts the elastic deformation of the boundary of a triangle under the complex function having the rule

$$f(z) = z^2 + 1.$$

The domain of f is the subset of $\mathbf{C}$ consisting of all elements $z \in \mathbf{C}$ such that they may be represented by the sides of the triangle ABC with vertices given by the ordered pairs $(0, 0)$, $(1, 0)$, (1.1). Now

$$\begin{aligned}
w &= u + iv \\
&= f(z) = z^2 + 1 \\
&= (x + iy)^2 + 1 \\
&= x^2 - y^2 + 1 + i2xy.
\end{aligned}$$

Since ordered pairs (x_0, y_0), (x_1, y_1) are equal if and only if $x_0 = x_1$ and $y_0 = y_1$, it follows that

$$u = x^2 - y^2 + 1, \ v = 2xy.$$

The function thus assigns the ordered pairs $(1, 0)$, $(2, 0)$, $(1, 2)$ in the codomain to $(0, 0)$, $(1, 0)$, $(1, 1)$ in the domain. The side AB of triangle ABC is the subset

$$\{(x, 0) : 0 \leq x \leq 1\}.$$

Substituting $y = 0$ into the expressions obtained for u and v now yields

$$u = x^2 + 1, \ v = 0.$$

Thus the image of $\{(x, 0) : 0 \leq x \leq \}$ is thus

$$\{(u, 0) : 1 \leq u \leq 2\}.$$

This is the side $A'B'$ in Figure 15.11. The image of the side BC is obtained by substituting $x = 1$ into the expressions for u and v, giving

$$u = 2 - y^2, \ v = 2y.$$

Elimination of y yields $v^2 = 8 - 4u$. Thus the image of

$$\{(1, y) : 0 \leq y \leq 1\}$$

is

$$\left\{\left(2 - \frac{v^2}{4}, v\right) : 0 \leq v \leq 2\right\}.$$

This is the curved side $B'C'$ depicted in Figure 15.11. Finally, the image of the side CA is obtained by substituting $y = x$ into the expressions for u and v, giving

$$u = 1, \ v = 2x^2.$$

Thus the image of

$$\{(x, x) : 1 \geq x \geq 0\}$$

is

$$\{(1, v) : 2 \geq v \geq 0\}.$$

This completes the triangle $A'B'C'$ of Figure 15.11.

   If the interior points of triangle ABC of Figure 15.11 are considered as well as the set of points comprising its boundary, then the domain of the function f becomes

$$\{(x, y) : 0 \leq x \leq 1 \text{ and } 0 \leq y \leq x\}.$$

It is not difficult to show that this set of interior points maps to the set of interior points of triangle $O'B'C'$. The deformation is not an affine transformation since straight lines are not preserved. Both triangles

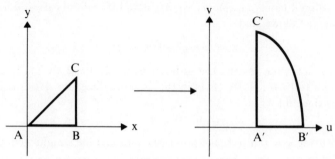

Fig. 15.11

are, however, homeomorphic to a disc, so such a deformation is clearly a permitted topological transformation.

Another function $f : \mathbf{C} \to \mathbf{C}$ is that having the rule

$$f(z) = \frac{1}{z}, \text{ where } z \neq 0$$

Here,

$$w = u + iv$$

$$= f(z) = \frac{1}{z}$$

$$= \frac{1}{x + iy}$$

$$= \frac{x}{x^2 + y^2} - i \frac{y}{x^2 + y^2}.$$

Thus,

$$u = \frac{x}{x^2 + y^2}, \quad v = -\frac{y}{x^2 + y^2}.$$

The effect of this function upon a subset of $\mathbf{R}^2$ can more readily be appreciated if polar coordinates are used. Thus, if the ordered pairs (x, y) are replaced by their corresponding polar representations (r, θ), where

$$x = r \cos \theta, \, y = r \sin \theta,$$

the image of each point (r, θ), will be the point $(1/r, -\theta)$. Each and every point z on a given ray from the origin is mapped to a corresponding point w which lies on the reflection of that ray across the line $\theta = 0$ or π and which is at a reciprocal distance from the origin. This deformation thus combines reflection across the x-axis with what is termed *inversion* with respect to the unit circle $x^2 + y^2 = 1$.

Figure 15.12 shows the transformation by f of the line AB, which is the subset

$$\{(x, x) : 1 \leqq x \leqq 2\}.$$

Substituting $y = x$ in the appropriate expressions for u and v yields

$$u = \frac{1}{2x}, v = -\frac{1}{2x}.$$

Also, the ordered pairs $(1, 1)$, $(2, 2)$ are mapped respectively to $(\frac{1}{2}, -\frac{1}{2})$, $(\frac{1}{4}, -\frac{1}{4})$. Hence the image of $\{(x, x) : 1 \leqq x \leqq 2\}$ is

$$\{(u, -u) : \tfrac{1}{2} \geqq u \geqq \tfrac{1}{4}\}.$$

In Figure 15.12, only one plane representing $\mathbf{R}^2$ has been used. The two components of the transformation have been indicated by also depicting the intermediate stage where AB is simply inverted with respect to the unit circle.

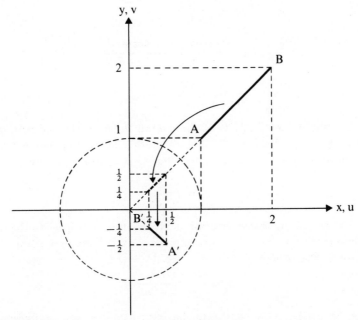

Fig. 15.12

Clearly, the function with the rule $f(z) = 1/z$ maps all points outside the unit circle to points inside the unit circle and vice versa, whilst at the same time mapping points above the x-axis to points below the x-axis and vice versa. The image of the point $(0, 0)$, if allowed, lies at an infinite distance from (0, 0) and is termed *the point at infinity*. Similarly the point at infinity, if included in the domain of the function, maps to (0, 0). This preserves the function as a bijection.

An extremely important class of functions $f: \mathbf{C} \to \mathbf{C}$ which combine translation, inversion, reflection, magnification and rotation is the class with the rule

$$f(z) = \frac{az+b}{cz+d},$$

where a, b, c, d are complex constants and $ad \neq bc$. By the following simple algebraic manipulation such functions may be rewritten in a form where the various components of the overall transformation can readily be interpreted:

$$f(z) = \frac{az+b}{cz+d}$$

$$= \frac{(acz+bc)/c^2}{z+d/c}$$

$$= \frac{(bc-ad)/c^2}{z+d/c} + \frac{acz+ad}{c^2(z+d/c)}$$

$$= \frac{(bc-ad)/c^2}{z+d/c} + \frac{a}{c}. \qquad \ldots (1)$$

In the step-by-step analysis of the overall function (1), the first step maps z to $z+(d/c)$. Since d/c is a constant, the corresponding transformation is a translation. The next step maps z_1 to $1/z_1$, where

$$z_1 = z+\frac{d}{c}.$$

The corresponding transformation is inversion with respect to the unit circle together with reflection across the x-axis. Next z_2 is mapped to $[(bc-ad)/c^2]z_2$, where

$$z_2 = \frac{1}{z_1} = \frac{1}{z+(d/c)}.$$

Since $(bc-ad)/c^2$ is a constant, the corresponding transformation is magnification (or contraction) together with rotation. This is easily seen in terms of polar coordinates, for, if the polar form of $(bc-ad)/c^2$ is (ρ, ϕ), then multiplication by (ρ, ϕ) magnifies and rotates each point (r, θ) representing z_2 so that the resulting distance from (0, 0) becomes ρr and the resulting angle is $\theta + \phi$. The final addition of the constant a/c is simply a further translation. It can be shown that this class of

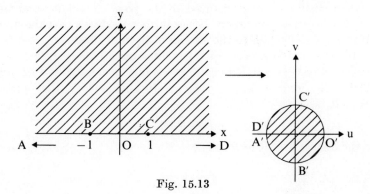

Fig. 15.13

functions, known as the class of *bilinear functions*, maps the set of straight lines and circles to the set of straight lines and circles.

Figure 15.13 depicts a mapping under the bilinear function $f : \mathbf{C} \rightarrow \mathbf{C}$ with the rule

$$f(z) = \frac{i-z}{i+z}.$$

Here,

$$w = u + iv$$

$$= f(z) = \frac{i-z}{i+z}$$

$$= \frac{i-(x+iy)}{i+(x+iy)}$$

$$= \frac{-x^2-y^2+1+i2x}{x^2+(1+y)^2}.$$

This gives

$$u = -\left(\frac{x^2+y^2-1}{x^2+(1+y)^2}\right), \quad v = \frac{2x}{x^2+(1+y)^2}$$

The image of the x-axis $\{(x, 0)\}$ is obtained by substitution of $y = 0$ in the expressions for u and v, giving

$$u = -\left(\frac{x^2-1}{x^2+1}\right), \quad v = \frac{2x}{x^2+1},$$

whence, on eliminating x,

$$u^2 + v^2 = 1,$$

which is the unit circle. By direct substitution into

$$f(z) = \frac{i-z}{i+z}$$

it is seen that $(-1, 0)$ is mapped to the point at infinity and that $(0, 1)$ is mapped to the origin. It can further be shown that the upper half plane, $y > 0$, maps to the interior of the unit circle.

If a function $f : X \to Y$ is a bijection, it has an *inverse* $f^{-1} : Y \to X$ which is also a bijection. For example, the bijection of Figure 15.4 (page 136) has the inverse shown in Figure 15.14. Similarly, the function $f : \mathbf{R} \to \mathbf{R}$, where

$$f(x) = 2x,$$

has the inverse $f^{-1} : \mathbf{R} \to \mathbf{R}$, where

$$f^{-1}(x) = \frac{x}{2}.$$

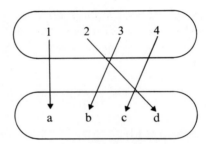

Fig. 15.14

If, however, a function f is only an injection, then, strictly, it does not have an inverse since there are elements of the codomain which are not images of any element in the domain. In reversing domain and codomain on forming the inverse, there would be elements with no image, and the result is thus not a function at all. If, however, the domain of the 'inverse' is restricted to the original image set, then the problem is circumvented. For example, the function with domain and codomain the set of non-negative integers $\mathbf{Z}_0^+$ and with the rule

$$f(x) = 2x$$

does not strictly have an inverse, since the images of the odd positive integers under the 'inverse' would not belong to $\mathbf{Z}_0^+$. Restricting the domain to the even non-negative integers circumvents the problem, but the resulting function is not the inverse of $f : \mathbf{Z}_0^+ \to \mathbf{Z}_0^+$ but of the bijection which maps elements of $\mathbf{Z}_0^+$ onto the even non-negative integers with the same rule, $f(x) = 2x$.

Again, the function $f : \mathbf{R} \to \mathbf{R}$ where

$$f(x) = x^2$$

does not have an inverse since not only does $f(\mathbf{R}) \neq \mathbf{R}$ but also the function is not one–one. In this case, for which $f^{-1}(x) = \sqrt{x}$, the element 1, for example, maps to both -1 and $+1$, and the images of negative numbers belong to $\mathbf{C}$ and not to $\mathbf{R}$. The function $f: \mathbf{R}_0^+ \to \mathbf{R}_0^+$ with the rule $f(x) = x^2$ does, however, have an inverse with

$$f^{-1}(x) = \sqrt{x},$$

since only the positive root is taken and the function is a bijection.

By the definition of inverse function, it follows that the inverse of the inverse of a function is the original function, that is,

$$(f^{-1})^{-1} = f.$$

Thus inversion of functions obeys the *involution law*. Further, there is a class of functions for which each function is its own inverse. This is the class of *identity functions*, $f: X \to X$, with the rule

$$f(x) = x.$$

The following functions are all examples of rigid or elastic transformations which do not necessarily involve tearing:

$f: \mathbf{R} \to \mathbf{R}$ with the rule $f(x) = ax+b$,

$f: \mathbf{R}_0^+ \to \mathbf{R}_0^+$ with the rule $f(x) = x^2$,

$f: \mathbf{R}^2 \to \mathbf{R}^2$ with the rule $f(x_0, x_1) = (x_0+a_0, x_1+a_1)$,

$f: \mathbf{R}^2 \to \mathbf{R}^2$ with the rule $f(x_0, x_1) = (ax_0, ax_1)$,

$f: \mathbf{C} \to \mathbf{C}$ with the rule $f(z) = \dfrac{az+b}{cz+d}$, where $ad \neq bc$.

Some functions, however, do involve tearing, and two examples are depicted in Figures 15.15 and 15.16. In Figure 15.15, which depicts an injection $f: \mathbf{R} \to \mathbf{R}$, the subset X of $\mathbf{R}$ which is the interval

$$\{x : a \leq x \leq b\}$$

has an image $f(X)$ consisting of two intervals of $\mathbf{R}$:

$$\{y : c \leq y \leq d\}$$

and

$$\{y : e \leq y \leq f\}$$

which are separated from each other. Thus X may be thought of as having been 'torn into two pieces' by the function f. In order to preserve the one–one correspondence, it is necessary to specify which of d and e is the image of x_0. In Figure 15.15, it has been carefully indicated that d is the image of x_0. This means that the boundary point

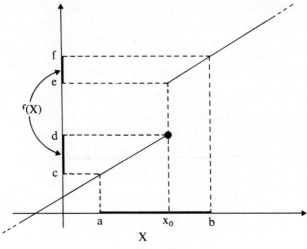

Fig. 15.15

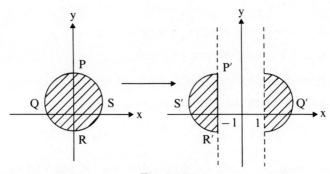

Fig. 15.16

e of the interval $\{y : e < y \leq f\}$ is not to be included. The subsets $\{x . a \leq x \leq b\}$ and $\{y : c \leq y \leq d\}$ are said to be *closed* since their boundary points are included. The subset $\{y : e < y \leq f\}$ is not closed, and further, if the other boundary point f is also excluded giving $\{y : e < y < f\}$, then it is said to be *open*. The 'intermediate' case where only one boundary point is included is sometimes said to be *half-open*. The concept of an open set plays a fundamental role in the definition of *continuity* in topology.

Figure 15.16 depicts an injection $f : \mathbf{R}^2 \rightarrow \mathbf{R}^2$ where the particular subset of $\mathbf{R}^2$ represented by the disc $PQRS$ is considered. The function maps (x, y) to $(-1-x, y)$ for non-negative x, and (x, y) to $(1-x, y)$ for negative x. The disc is thus 'torn in half' by the function as well as

being reflected. One–one correspondence is preserved by mapping all points lying on PR to points on $P'R'$. Thus the left-hand half-disc $P'R'S'$ is a closed set, since the whole of its boundary is included, whilst the right-hand half-disc is not closed, only part of its boundary being included.

It has been seen in this chapter that rigid and elastic transformations can be defined by appropriate functions, and that tearing is intuitively associated with functions that are discontinuous. So long as the domains and codomains of functions are either finite sets or subsets of $\mathbf{R}$ or $\mathbf{R}^2$ (or even $\mathbf{R}^3$) then a pictorial representation in one form or another is both practicable and useful. In order to arrive at a more general understanding of continuity, however, it is necessary to develop the concepts so far introduced in a more abstract context.

16

Metric Spaces

Distance in $\mathbf{R}^n$—definition of metric—neighbourhoods—continuity in terms of neighbourhoods—complete system of neighbourhoods—requirement for proof of non-continuity—functional relationships between δ and ε—limitations of metric.

In Chapter 13 it was seen that for three-dimensional Euclidean space, the distance between two points P_0, P_1 satisfies the conditions:

$d(P_0, P_1) \geqq 0$;

$d(P_0, P_1) = 0$ if and only if $P_0 = P_1$;

$d(P_0, P_1) = d(P_1, P_0)$;

$d(P_0, P_1) + d(P_1, P_2) \geqq d(P_0, P_2)$.

Generally, for x and y belonging to $\mathbf{R}^n$, the distance $d(x, y)$ may be defined as
$$d(x, y) = \sqrt{[(x_1 - y_1)^2 + \ldots + (x_n - y_n)^2]},$$
where the n-tuples $(x_1, \ldots, x_n)$, $(y_1, \ldots, y_n)$ are the coordinates of x and y respectively. For any points x, y, z of $\mathbf{R}^n$, distance satisfies the conditions:

$d(x, y) \geqq 0$;

$d(x, y) = 0$ if and only if $x = y$;

$d(x, y) = d(y, x)$;

$d(x, y) + d(y, z) \geqq d(x, z)$.

It is not, however, necessary to confine the concept of distance to n-dimensional Euclidean space. If X is any non-empty set for which a non-negative real number $d(x, y)$ is defined for all elements $x, y \in X$ satisfying the four conditions above, then d is termed a *metric* and the set X together with the metric d is termed a *metric space*. The metric d is a function $X \times X \to \mathbf{R}_0^+$. A simple example of a metric space is provided by any non-empty set X together with a metric d defined by

$$d(x, y) = \begin{cases} 0 \text{ if } x = y \\ a > 0 \text{ if } x \neq y \end{cases}$$

150

for all $x, y \in X$. It can easily be seen that the four conditions for a metric are satisfied by d as defined above.

Another example providing a set of metric spaces is the set of all n-figure binary numbers with the metric d defined as the number of changes of digits required in going from one number to another. Thus if in the set of five-figure binary numbers $x = 01101$ and $y = 10111$, then $d(x, y) = 3$, since there is a change in each of the first, second and fourth digit places (reading from left to right).

In Chapter 13, it was seen that for any point $x_0 \in \mathbf{R}$, the set of all points x of $\mathbf{R}$ satisfying

$$d(x_0, x) < \delta$$

(where d is the usual metric in $\mathbf{R}$) is termed a *neighbourhood* (or *open ball*) of x_0. Clearly, a different neighbourhood of x_0 is defined for each chosen value of δ. The subset satisfying $d(x_0, x) < \rho$ is termed the neighbourhood of x_0 of *radius* ρ and denoted by $N(x_0; \rho)$.

Figure 16.1 depicts a point x_0 belonging to a subset X of $\mathbf{R}^2$, where X is the set of all points within and on the continuous closed curve C. The neighbourhood of x_0 in $\mathbf{R}^2$ of radius ρ is represented by the circle with centre x_0 and radius r, excluding the circumference. The neighbourhood of x_0 in X of radius ρ is represented by that part of the interior of the circle which is shaded, including interior points of the circle belonging to C. Clearly,

$$N(x_0 \in X; \rho) = X \cap N(x_0 \in \mathbf{R}^2; \rho),$$

and a similar relationship will hold for any point $x \in X \subset \mathbf{R}^2$.

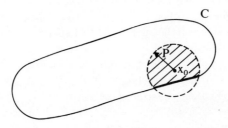

Fig. 16.1

In the case of the non-empty set X and the metric d defined by

$$d(x, y) = \begin{cases} 0 \text{ if } x = Y \\ a > 0 \text{ if } x \neq y, \end{cases}$$

the neighbourhood in X of any point $x \in X$ of radius a consists of the single point x itself, since any point other than x is defined to have distance a from x. In the case of the set of five-figure binary numbers with d defined as the number of changes of digits in going from one number to another, the neighbourhood of 01101 of radius 3 is the subset

$$\{01101, \ 11101, \ 00101, \ 01001, \ 01111, \ 01100, \ 10101, \ 11001,$$
$$11111, \ 11100, \ 00001, \ 00111, \ 00100, 01011, 01000, 01110\}.$$

This is the subset of all five-figure binary numbers differing by zero, one, or two digits from 01101. The concept of a neighbourhood of given radius may thus be generalised to sets other than subsets of $\mathbf{R}^n$.

The definition of continuity of functions, given in Chapter 13, may now be re-expressed in terms of neighbourhoods where functions are mapping one metric space to another. If X, Y are two metric spaces with metrics d, e respectively, and if $x_0 \in X$, then a function $f : X \to Y$ is said to be *continuous at* x_0 if, for each neighbourhood of $f(x_0)$ in Y, there is some neighbourhood of x_0 in X whose image is in the neighbourhood of $f(x_0)$ in Y. Symbolically, if for every radius ε there is a radius δ such that

$$f(N(x \in X ; \delta)) \subseteq N(f(x) \in Y ; \varepsilon),$$

then f is continuous at x_0. If this holds for all $x \in X$, then the function is continuous at each and every point of its domain and is said simply to be *continuous*. Thus 'continuous' means 'continuous everywhere'.

The term *open ball* for a neighbourhood of given radius arises from a consideration of neighbourhoods in $\mathbf{R}^3$. Such a neighbourhood in $\mathbf{R}^3$ has a radius defining its boundary and consists of the set of all the points *within* this boundary. Such a set is an open set, and a given point of $\mathbf{R}^3$ together with the given radius defines the interior of a sphere. Hence the term 'open ball' (rather than *open sphere*) is used to emphasise that it is not the surface of the sphere which is being considered. A neighbourhood of a general nature is simply a subset of points containing a given point together with some open ball about that point. The set of all neighbourhoods of a given point $x \in X$ is termed the *complete system of neighbourhoods* of x.

An example of a function which is continuous over a certain subset of $\mathbf{R}^2$ but not continuous everywhere was depicted in Figure 15.16 (page 148). The same function is depicted in Figure 16.2 where $X = \{(x, y) : x < 0\}$, and $Y = \{(x, y) : x \geqq 0\}$. The images $f(X)$, $f(Y)$ are as depicted in the figure. If $a = (0, a)$ is some point of the subset $\{(0, y)\}$, then $a' = f(a) = (-1, a)$ is a point of the subset $\{(-1, y)\}$. If $0 < \varepsilon < 2$, then the neighbourhood of a' in $\mathbf{R}^2$, $N(a' ; \varepsilon)$ contains no points of $f(X)$. However, all neighbourhoods of a in $\mathbf{R}^2$ do contain points belonging to X. There is thus no δ corresponding to ε such that

$$f(N(a ; \delta)) \subseteq N(a' ; \varepsilon),$$

and the function is therefore not continuous at a. Since a was taken as an arbitrary point of $\{(0, y)\}$ it follows that f is nowhere continuous on $\{(0, y)\}$. If, however, any point of X or Y is chosen not belonging to $\{(0, y)\}$, there will always be some δ, however small, satisfying the continuity condition. The function is thus continuous everywhere in $\mathbf{R}$ except for the subset $\{(0, y)\}$.

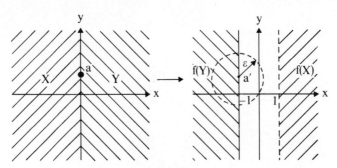

Fig. 16.2

To prove that a function is **not** continuous it is only necessary to find some ε for which no δ exists satisfying the continuity condition. Proving that a function **is** continuous is more difficult because it is necessary to establish that some δ exists for every point of its domain and every possible choice of ε. This may sometimes be done by finding a suitable functional relationship between δ and ε. In certain cases it is possible to take δ equal to ε, the simplest example of which is the identity function $f : \mathbf{R}^n \to \mathbf{R}^n$ for which $f(x) = x$. Other examples are provided by the various rigid transformations. In the case of similarity transformations δ may be taken equal to ε if the transformation is one of contraction. But, if the function increases distances by a factor a greater than unity it is necessary to take $\delta = \varepsilon/a$.

An example of finding an appropriate functional relationship between δ and ε is provided by consideration of the continuity of the radial projection $f : \mathbf{R}^3 - c \to S$, which projects all points of $\mathbf{R}^3$ except c on to the surface S of a sphere of unit radius and centre c. This is depicted in Figure 16.3. The distance between any two points exterior to the sphere is shrunk by the function f, and the distance between any two points on the surface S remains unchanged. Hence, f is continuous if its domain is restricted to elements x of $\mathbf{R}^3$ for which

$$d(x, c) \geqq 1.$$

For pairs of points inside S, however, the function f magnifies distances more and more as the points become closer and closer to c, and con-

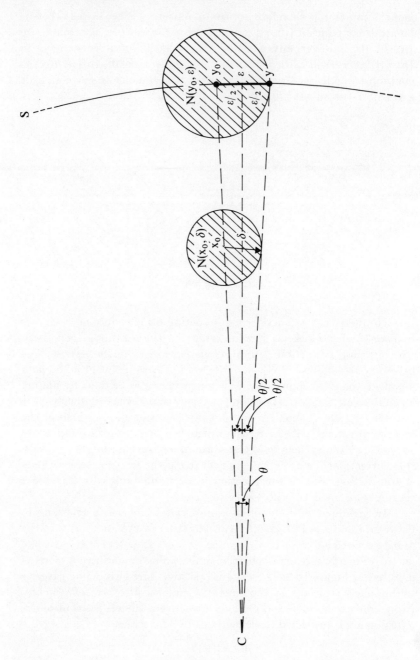

Fig. 16.3

tinuity cannot be established by the argument applying to exterior points.

The continuity of f may be established, however, as follows. Let x_0 be some point of $\mathbf{R}^3$ inside S, and let y_0 be the corresponding image $f(x_0)$ on S. Let a neighbourhood of y_0 be denoted by $N(y_0; \varepsilon)$ and let y be some point on the intersection of the sphere, having centre y_0 and radius ε, with the surface S. Now, let S be the perpendicular distance from x_0 to the radius $c\text{-}y$. The continuity of f at x_0 follows, since it is clear that for any ε the chosen δ will satisfy $f(N(x; \delta)) \subseteq N(y; \varepsilon)$, and since x_0 is an arbitrary point inside S, continuity follows for all points inside S, and hence also for all points belonging to $\mathbf{R}^3 - c$. The functional relationship between δ and ε is given by:

$$\delta = d(x_0, c) \cdot \sin \theta$$

$$= 2 \cdot d(x_0, c) \cdot \sin \frac{\theta}{2} \cos \frac{\theta}{2}$$

$$= 2 \cdot d(x_0, c) \cdot \frac{\varepsilon}{2} \sqrt{\left(1 - \frac{\varepsilon^2}{4}\right)}$$

$$= d(x_0, c) \cdot \sqrt{\left(\varepsilon^2 - \frac{\varepsilon^4}{4}\right)}$$

Since topology is concerned with the study of properties which remain invariant under certain permitted transformations, and since these transformations are defined as *continuous functions having continuous inverses*, the establishment of continuity by the methods discussed in this chapter is an important stage in the investigation of topological invariants. However, these methods are dependent upon the existence of a metric, and so can apply to metric spaces only. A still more general approach is needed, and this will require continuity to be defined without recourse to any metric so that sets representing non-metric spaces may be embraced. It has already been seen that neighbourhoods are a particular kind of *open set*, and it is this concept of *open set* which provides the basis for a definition of continuity not requiring any reference to a metric. Such a general approach will be developed in the next chapter.

17

Topological Spaces

Concept of open set—definition of a topology on a set—topological space—examples of topological spaces—open and closed sets—redefining neighbourhood—metrizable topological spaces—closure—interior, exterior, boundary—continuity in terms of open sets—homeomorphic topological spaces—connected and disconnected spaces—covering—compactness—completeness: not a topological property—completeness of the real numbers—topology, the starting point of real analysis.

In Chapter 16 a formal definition of continuity in the context of metric spaces was given which depends ultimately on the existence of the appropriate metrics. If the concept of distance is dropped altogether, but the concept of open set is retained, then a new concept arises, namely that of a *topological space*.

If X is any non-empty set and $\mathscr{T}$ a collection of subsets of X including the empty set $\varnothing$ and the set X itself, then the collection $\mathscr{T}$ is termed a *topology on X* provided that it satisfies each of the following conditions:

1. The union of any number of the subsets of X which are in $\mathscr{T}$ must also be in $\mathscr{T}$.
2. The intersection of any two of the subsets of X which are in $\mathscr{T}$ must also be in $\mathscr{T}$.

These two conditions may be written symbolically as:

1. If $A_i \in \mathscr{T}$ for all $i \in M$, then
$$\bigcup_{i \in M} A_i \in \mathscr{T}.$$
2. If $B, C \in \mathscr{T}$, then
$$B \cap C \in \mathscr{T}.$$

The various members of the collection $\mathscr{T}$ are said to be $\mathscr{T}$-*open* (or simply *open*). Thus the statement that a given set B is a $\mathscr{T}$-open set is simply equivalent to the statement that B belongs to the collection $\mathscr{T}$. The set X together with the collection $\mathscr{T}$ comprises a *topological space*.

A first example of a topological space is provided by the set

$$X = \{a, b, c, d\}$$

together with the collection of subsets of X,

$$\mathscr{T} = \{\varnothing, \{a\}, \{a, b\}, \{a, b, d\}, X\}.$$

The union of any number of these subsets belongs to $\mathscr{T}$. This is ensured in this instance by the hierarchical relationship

$$\varnothing \subset \{a\} \subset \{a, b\} \subset \{a, b, d\} \subset X,$$

which also ensures that the intersection of any number of these subsets (and hence of any two of them) belongs to $\mathscr{T}$. The collection of subsets

$$\{\varnothing, \{a\}, \{a, b\}, \{b, c, d\}, X\}$$

is not, however, a topology on X since the intersection, $\{a, b\} \cap \{b, c, d\} = \{b\}$ gives a subset not belonging to the collection.

A second example of a topological space is provided by the set of integers $\mathbf{Z}$ and the collection of subsets of $\mathbf{Z}$,

$$\mathscr{T} = \{\varnothing, \mathbf{Z}_{odd}, \mathbf{Z}_{even}, \mathbf{Z}\}.$$

In this case, there is no longer the same hierarchical structure as in the previous example, but two 'parallel' relationships

$$\varnothing \subset \mathbf{Z}_{odd} \subset \mathbf{Z},$$

$$\varnothing \subset \mathbf{Z}_{even} \subset \mathbf{Z},$$

the middle elements of which are disjoint subsets of $\mathbf{Z}$. This type of hierarchy, however, still guarantees that the two conditions are satisfied.

A third example of a topological space is provided by the set of real numbers $\mathbf{R}$ together with the collection of all open subsets of $\mathbf{R}$, that is, every continuous open interval of $\mathbf{R}$ including $\mathbf{R}$ itself and the empty set $\varnothing$. This particular collection is a topology on $\mathbf{R}$ termed the *usual topology* on $\mathbf{R}$. Similar collections form the usual topologies on $\mathbf{R}^n$ for higher order spaces.

Any non-empty set X has two trivial topologies. One of these, termed the *indiscrete topology* on X, consists of the two subsets $\varnothing$ and X only. Together with X, it forms an *indiscrete topological space*. The other, termed the *discrete topology* on X, consists of all possible subsets of X (including $\varnothing$ and X), and together with X forms a *discrete topological space*. Thus, if as before

$$X = \{a, b, c, d\},$$

the indiscrete topology on X is the pair of subsets

$$\{\varnothing, \{a, b, c, d\}\},$$

neither of which is a proper subset of X, and the discrete topology on X is the power set

$$\mathscr{P}(X) = \{\varnothing, \{a\}, \{b\}, \{c\}, \{d\}, \{a, b\}, \{a, c\}, \{a, d\}, \{b, c\}, \{b, d\},$$
$$\{c, d\}, \{a, b, c\}, \{a, b, d\}, \{a, c, d\}, \{b, c, d\}, \{a, b, c, d\}\}.$$

The intersection of any number of topologies on a given set X is also a topology on X, since it is easily seen that the intersection of any two topologies on X satisfies the two prescribed conditions, and this is readily generalized to the intersection of any number of topologes. The union of topologies on a given set X need not, however, be a topology on X. A demonstration of this is provided by the set

$$X = \{a, b, c\}$$

and the two topologies on X,

$$\mathscr{T}_1 = \{\varnothing, \{a\}, X\},$$
$$\mathscr{T}_2 = \{\varnothing, \{b\}, X\}.$$

The union,

$$\mathscr{T}_1 \cup \mathscr{T}_2 = \{\varnothing, \{a\}, \{b\}, X\},$$

violates the second prescribed condition since $\{a\}$ and $\{b\}$ both belong to $\mathscr{T}_1 \cup \mathscr{T}_2$ but $\{a\} \cup \{b\} = \{a, b\}$ does not.

An alternative definition of a topological space may be given in terms of $\mathscr{T}$-*closed* subsets. A subset $A \subseteq X$ is $\mathscr{T}$-*closed* (or simply *closed*) if its relative complement in X, namely $X - A$, is $\mathscr{T}$-open. A collection $\mathscr{K}$ of such subsets of X which includes X and $\varnothing$ is a topology on X if it satisfies the two conditions:

1. The intersection of any number of subsets of X which are in $\mathscr{K}$ must also be in $\mathscr{K}$.
2. The union of any two of the subsets of X which are in $\mathscr{K}$ must also be in $\mathscr{K}$.

Symbolically, these conditions become:

1. If $A_i \in \mathscr{K}$ for all $i \in M$, then

$$\bigcap_{i \in M} A_i \in \mathscr{K}.$$

2. If $B, C \in \mathscr{K}$, then

$$B \cup C \in \mathscr{K}.$$

If $X = \{a, b, c, d\}$ as before, the relative complements in X of the members of the topology on X consisting of the open subsets

$$\varnothing, \{a\}, \{a, b\}, \{a, b.\ d\}, X$$

are the closed subsets

$$X, \{b, c, d\}, \{c, d\}, \{c\}, \varnothing$$

respectively. These closed subsets form a collection which is a closed set topology on X since, as can readily be seen, the two prescribed conditions are satisfied. In this instance, however, the collection of subsets

$$\{\varnothing, \{c\}, \{c, d\}, \{b, c, d\}, X\}$$

also satisfies the two conditions for open sets, and the collection of relative complements

$$\{X, \{a, b, d\}, \{a, b\}, \{a\}, \varnothing\},$$

originally considered as a $\mathscr{T}$-open collection, also satisfies the two conditions for closed sets. A given subset may therefore be open in one topology and closed in another. Clearly, the empty set $\varnothing$ and the set X itself, whatever its elements, are always both open and closed. Also, given a topology $\mathscr{T}$ on a set X, there may be subsets which are both $\mathscr{T}$-open and $\mathscr{T}$-closed, and also subsets which are neither $\mathscr{T}$-open nor $\mathscr{T}$-closed.

The definitions of open and closed sets given in the context of topological spaces is more general than the intuitive definition arrived at by considering subsets of $\mathbf{R}^3$, and defining a subset as *open* if it contains no points of its boundary and *closed* if it contains all points of its boundary. Metric spaces are, however, examples of topological spaces, since for any metric space X with metric d, the set X together with the collection of all open subsets of X gives an associated topological space. There is thus a need for a link between the concept of neighbourhood in a metric space and the concept of open set in a topological space. This is provided by redefining neighbourhood in a purely topological context: if X is a set which, together with some topology $\mathscr{T}$ on X, is a topological space, then any subset N of X is a *neighbourhood* of an element $x \in X$ if it includes an open set containing x. This particular redefinition of neighbourhood ensures that its interpretations in a metric context and in a topological context are consistent.

A given set may give rise to several distinct topological spaces. Thus, the set $X = \{x, y\}$ has four distinct topologies on it, namely

$$\mathcal{T}_1 = \{\varnothing, X\},$$
$$\mathcal{T}_2 = \{\varnothing, \{x\}, X\},$$
$$\mathcal{T}_3 = \{\varnothing, \{y\}, X\},$$
$$\mathcal{T}_4 = \{\varnothing, \{x\}, \{y\}, X\}.$$

Each of $\mathcal{T}_1, \ldots, \mathcal{T}_4$ satisfy the two conditions for a topology, and hence there are four distinct topological spaces $(X, \mathcal{T}_1)$, $(X, \mathcal{T}_2)$, $(X, \mathcal{T}_3)$, $(X, \mathcal{T}_4)$ each arising out of the same set X. Any metric d defined on the set X must satisfy the three conditions:

$d(x, x) = 0;$

$d(y, y) = 0;$

$d(x, y) = d(y, x) = a > 0.$

A topological space associated with the metric space (X, d) is obtained from the collection of all open subsets of the metric space, that is, the collection of all possible unions of neighbourhoods of elements of X. Consider the neighbourhoods

$$N(x\,;a/2), \; N(y\,;a/2).$$

These neighbourhoods consist of the two subsets having the single elements x and y respectively. The unions of these give as the associated topology, the set

$$\{\varnothing, \{x\}, \{y\}, X\},$$

which is the topology $\mathcal{T}_4$ above. Since, an arbitrary metric was chosen, in that a was arbitrary, $\mathcal{T}_4$ is the only topology arising from the metric d and hence $(X, \mathcal{T}_4)$ is the only possible associated topology. When a topological space can be associated with a metric space in this way, it is said to be *metrizable*. Thus $(X, \mathcal{T}_4)$ is metrizable, but $(X, \mathcal{T}_1)$, $(X, \mathcal{T}_2)$, $(X, \mathcal{T}_3)$ are not metrizable.

Metrizable topological spaces are examples of an important class of topological spaces termed *Hausdorff spaces*. Hausdorff spaces are topological spaces satisfying the condition that for each pair of distinct points x, y of a set X with topology $\mathcal{T}$, there are neighbourhoods N_x and N_y of x and y respectively such that their intersection is empty.

The *closure* of a subset A of a topological space X will be denoted by A^- and is defined to be the intersection of all closed subsets of X containing A. That is:

$$A^- = \bigcap \{F : F \subseteq X, F \text{ is closed, and } F \supseteq A\}.$$

Clearly, since A^- is the intersection of a number of closed sets, it must itself be closed. Further, it is immediately seen that A^- is the smallest closed subset of X containing A, so that, if F is a closed subset of X containing A, then $A \subseteq A^- \subseteq F$. A itself will be a closed subset of X if it is equal to its closure.

Returning to an earlier example where $X = \{a, b, c, d\}$ and the collection

$$\{X, \{b, c, d\}, \{c, d\}, \{c\}, \varnothing\}$$

is a collection of closed subsets of X, it follows that $\{b\}^- = \{b, c, d\}$, $\{a, b\}^- = X$ and $\{c, d\}^- = \{c, d\}$.

The association of a new subset A^- with each subset A of a topological space X satisfies the following five properties:

$$\varnothing^- = \varnothing,$$

$$X^- = X,$$

$$A \subseteq A^- \quad \text{for every } A \text{ of } X,$$

$$(A \cup B)^- = A^- \cup B^- \quad \text{for every } A, B \text{ of } X,$$

$$(A^-)^- = A^- \quad \text{for every } A \text{ of } X.$$

These properties may be used as a set of axioms for what may be defined as a *closure space*. There is then a one–one correspondence between the collection of closure spaces and the collection of topological spaces.

Another important subset which may be associated with a subset A of a topological space X is the *interior* of A. A point $a \in A$ is called an *interior point* of A if it belongs to an open subset of A, that is, if it has a neighbourhood contained in A. The set of all interior points of A is called the *interior* of A and is denoted by int (A). Clearly, the interior of A is the union of all open subsets of A, and is itself the largest open subset of A. That is, if G is an open subset of A, then

$$G \subseteq \text{int} (A) \subseteq A.$$

The set A will be an open subset of X if it is equal to its own interior.

The *exterior* of A is the set of all *exterior points* of A, that is, of all points which have a neighbourhood having no points in common with A. It is denoted by ext (A), and is the interior of the complement of A in X:

$$\text{ext} (A) = \text{int} (X - A).$$

The *boundary* of A, denoted by bdy (A), is the set of points not belonging either to int (A) or to ext (A). This means that a point x is in the boundary of A if it is in both the closure of A and the closure of the complement of A in X:

$$\text{bdy} (A) = A^- \cap (X - A)^-,$$

and it follows that

$$A^- = \text{int} (A) \cup \text{bdy} (A).$$

Since the boundary of A is the intersection of two closed sets, it must itself be closed.

Consider again, for example, the set $X = \{a, b, c, d\}$ with topology $\mathcal{T}$ given by the collection

$$\{\varnothing, \{a\}, \{a, b\}, \{a, b, d\}, X\},$$

and let the subset A of X be the set $\{a, b, c\}$. The points a and b are interior points of A since $a, b \in \{a, b\} \subset A$ and $\{a, b\}$ is an open set. The point c is not, however, an interior point of A, hence

$$\text{int } (A) = \{a, b\}.$$

The complement of A in X, $X - A$, is $\{d\}$, and int $(\{d\}) = \varnothing$, since d is not an interior point of $X - A$. Hence

$$\text{ext } (A) = \varnothing.$$

Accordingly,

$$\text{bdy } (A) = \{c, d\}.$$

As a second example, consider the set of real numbers $\mathbf{R}$ and four subsets of $\mathbf{R}$ given by the intervals $\{x \in \mathbf{R} : a \leqq x \leqq b\}$, $\{x \in \mathbf{R} : a < x < b\}$, $\{x \in \mathbf{R} : a < x \leqq b\}$, $\{x \in \mathbf{R} : a \leqq x < b\}$. These subsets of R may be denoted by $[a, b]$, $]a, b[$, $]a, b]$, $[a, b[$ respectively. Each of these has as interior the set $\{x \in \mathbf{R} : a < x < b\} =]a, b[$, and as boundary the set $\{a, b\}$.

For a third example, let x_0 be a point in the space $\mathbf{R}^3$, and consider the set S defined by

$$S = \{x \in \mathbf{R}^3 : d(x_0, x) = 1\},$$

The set S is closed in $\mathbf{R}^3$ and $(\mathbf{R}^3 - S)^- = \mathbf{R}^3$. Hence

$$\text{bdy } (S) = S^- \cap (\mathbf{R}^3 - S)^- = S \cap \mathbf{R}^3 = S.$$

Thus, when considered as a subset of $\mathbf{R}^3$, the set S is its own boundary. However, suppose that S is to be considered as a subset of itself. The complement of S in S is the null set $\varnothing$, hence the boundary of S is empty.

These formal definitions of interior, exterior and boundary are precise ways of putting into rigorous and more general form what is intuitively understood by the same terms when used in the context of some interval, area or volume in ordinary one-, two- or three-dimensional Euclidean space respectively.

The continuity of a function has been defined earlier (in Chapter 13, p. 123) in terms of neighbourhoods. Since all neighbourhoods are open sets, the continuity of a function may be defined alternatively, and more generally, in terms of open sets. Let $(X, \mathscr{T}_1)$ and $(Y, \mathscr{T}_2)$ be two topological spaces. A function

$$f : (X, \mathscr{T}_1) \to (Y, \mathscr{T}_2)$$

is said to be *continuous* if the inverse image of every $\mathscr{T}_2$-open subset of Y is a $\mathscr{T}_1$-open subset of X, that is if $A \in \mathscr{T}_2$ implies that $f^{-1}(A) \in \mathscr{T}_1$.

A further alternative definition in terms of closed sets is also possible since the complement in X of the inverse image of a subset $A \subseteq Y$ is the same as the inverse image of the complement of A in Y. Hence, a function

$$f : (X, \mathscr{T}_1) \to (Y, \mathscr{T}_2)$$

is continuous if the inverse image of every $\mathscr{T}_2$-closed subset of Y is a T_1-closed subset of X.

It is important to notice that these two definitions of the continuity of a function are expressed in terms of *inverse* images. There are, for example, numerous cases where a function

$$f : (X, \mathscr{T}_1) \to (Y, \mathscr{T}_2)$$

has the property that the image $f(A)$ of every open subset A of X is an open subset of Y, and yet the function is not continuous.

The concept of *homeomorphism* can now be defined in terms of topological spaces. Two topological spaces $(X, \mathscr{T}_1)$, $(Y, \mathscr{T}_2)$ are *homeomorphic* if there exists a bijection $f : X \to Y$ such that a subset A of X is $\mathscr{T}_1$-open if and only if $f(A)$ is $\mathscr{T}_2$-open.

Two non-empty subsets A, B of a topological space $(X, \mathscr{T})$ are said to be *separated* if each is disjoint from the closure of the other, that is, if

$$A^- \cap B = A \cap B^- = \varnothing.$$

The space $(X, \mathscr{T})$ may, alternatively, be said to have a separation if A and B are both open, and

$$A \cup B = X,$$

whilst

$$A \cap B = \varnothing.$$

For example, consider the subset of **R** defined by the interval $[a, b]$. Let c be an interior point of this interval, that is, let $a < c < b$, so that c divides $[a, b]$ into two parts. If the two parts are taken to be the closed sub-intervals $[a, c]$, $[c, b]$, it will immediately be seen that these are not disjoint since

$$[a, c] \cap [c, b] = \{c\} \neq \varnothing.$$

Suppose, however, that c is removed from one of the parts, say the second, so that the division is into the closed sub-interval $[a, c]$ and the half-open sub-interval $]c, b]$. The two parts are now disjoint

$$[a, c] \cap]c, b] = \varnothing,$$

and their union is the whole subset $[a, b]$ of $\mathbf{R}$. However, the closure of $]c, b]$ is $[c, b]$ and so

$$[a, c] \cap]c, b]^- \neq \varnothing.$$

If $[a, c]$ and $[c, b]$ are to be separated, it is necessary to remove c from both sub-intervals so that each becomes a half-open set. Thus, $[a, c[$ and $]c, b]$ are separated and the two conditions

$$[a, c[^- \cap]c, b] = [a, c] \cap]c, b] = \varnothing,$$

$$[a, c[\cap]c, b]^- = [a, c[\cap [c, b] = \varnothing$$

are satisfied. In order to obtain this separation of the interval $[a, b]$ of $\mathbf{R}$ it has been necessary to remove an interior point c entirely. Thus, it is the complement of c in $[a, b]$, namely $[a, b] - \{c\}$, and not $[a, b]$ which has the separation. It is not possible to obtain such a separation of any continuous interval of $\mathbf{R}$, nor of $\mathbf{R}$ itself, without removing some interior point. $\mathbf{R}$, thus, has no separation and is therefore said to be *connected*. Any topological space which is not connected is said to be *disconnected*. It should readily be appreciated that the formal definition of a connected space just given is equivalent in the case of surfaces to the more intuitive definition given at the beginning of Chapter 5.

An alternative definition of connectivity may be given in terms of subsets of a topological space which are at the same time both open and closed. Thus a topological space $(X, \mathscr{T})$ is connected if the only two subsets of X that are at the same time open and closed are the set X itself and the null set $\varnothing$.

Connectedness is a topological property, thus, if

$$f : (X, \mathscr{T}_1) \rightarrow (Y, \mathscr{T}_2)$$

is a permitted topological transformation, then $f(X)$ is connected if X is connected. Connectedness has a number of important applications. For example, it is the basis of the *intermediate value theorem*, which states that if an interval $[a, b]$ of $\mathbf{R}$ is mapped by a continuous function f into $\mathbf{R}$, then each value between $f(a)$ and $f(b)$ must be the image of at least one point in $[a, b]$. This can be seen intuitively in Figure 17.1. No matter what value y, such that $f(a) < y < f(b)$, is chosen, it must be the image of some $x \in [a, b]$.

In Chapter 10, the concept of a *fixed point theorem* was introduced. Formal proofs of fixed point theorems also depend upon the concept of a connected topological space.

A collection $\mathscr{C}$ of subsets of a topological space $(X, \mathscr{T})$ is called a *covering* of a subset $A \subseteq X$ if the union of the members of $\mathscr{C}$ contains A. Formally, $\mathscr{C}$ is a covering of A if

$$A \subseteq \bigcup \{C : C \in \mathscr{C}\}.$$

If each $C \in \mathscr{C}$ is an open set, then the collection $\mathscr{C}$ is called an *open covering* of A. If the collection $\mathscr{C}$ is finite, then $\mathscr{C}$ is a *finite covering* of A.

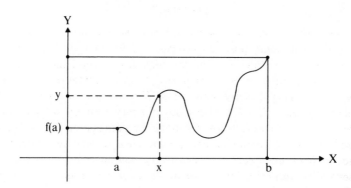

Fig. 17.1

A subset A of a topological space $(X, \mathscr{T})$ is said to be *compact* if every open covering of A contains a finite covering of A, that is, if from any infinite collection of open subsets whose union contains A it is possible to select a finite subcollection whose union also contains A. Again consider the set of real numbers **R**, and let A be the closed interval $[a, b]$ of **R**.
Let

$$\mathscr{C} = \{C_i : i \in M\}$$

be a collection of open intervals which is a covering of A, that is

$$A \subseteq \bigcup_i C_i.$$

An important property of this interval $[a, b]$ is that $\mathscr{C}$ does in fact contain a finite subcollection, also a covering of A. Suppose, however, that the interval is infinite, say $[a, \infty[$. In this case, no finite subcollection of a covering $\mathscr{C}$ can be a covering of A. Thus, any closed finite interval of **R** is compact, but open or infinite intervals are not compact. It is also clear that **R** itself is not compact. Any subset of **R** is compact if and only if it is closed and bounded (that is, contained in some interval $]-N, N[$ for sufficiently large N). An important property of Hausdorff spaces, of which the real line is an example, is that in a compact space any subset which is closed is also compact *and conversely*; thus closure and compactness may be thought of as equivalent for this particular class of topological spaces.

The final property of spaces to be considered here is the property of *completeness*. This, however, is not a topological property, but it is included in this chapter because of its importance in relation to the set of real numbers **R**. Let X be a metric space with metric d, and let

$$\{x_n : n \in \mathbf{N}\}$$

be a sequence in X. This is termed a *Cauchy sequence* if, for every $\varepsilon > 0$, there is some positive integer $n_0 \in \mathbf{N}$ such that the distance between every two members of the sequence beyond the n_0'th is less than ε, that is, if $n, m > n_0$, then $d(x_n, x_m) < \varepsilon$. The metric space X is *complete* if every Cauchy sequence in X converges to a member of X. Similarly, if $A \subset X$, A is complete if every Cauchy sequence in A converges to a member of A. Clearly, the concept of completeness is related to that of compactness. In fact, every compact subset of a metric space is complete, but not every complete subset is compact. The set of real numbers **R**, for example, is complete but not compact.

To see that completeness is not a topological property, consider the subset $X \subset \mathbf{R}$ defined by the interval $]-1, 1[$. X is not complete since the Cauchy sequence in X given by

$$1 - \frac{1}{n}, \; n \in \mathbf{N}$$

converges to 1, and 1 is not in X. Now the function $f : \mathbf{R} \to X$ defined by

$$f(x) = \frac{x}{|x| + 1}$$

is one–one, continuous and has a continuous inverse, and hence preserves topological properties. Neither **R** nor X, for example, are compact. **R** is, however, complete, whilst X has been shown not to be complete. Completeness therefore cannot be a topological property.

The concept of *irrational numbers*, that is, numbers, such as $\sqrt{2}$, which cannot be expressed in the form m/n where m and n are integers is a familiar concept which has been known since the days of Pythagoras. The more general concept of *non-algebraic numbers*, such as π, which are not roots of any polynomial equation with integer coefficients, is more recent but has certainly been a familiar concept for more than a century. The completeness of the real numbers **R** can be thought of as expressing formally the intuitive idea that it is now no longer possible to discover a new kind of real number. Another way of putting it is to say that the real line is completely filled by the numbers which are now known. There is thus a hierarchy of numbers commencing with the set of integers Z, and proceeding through the set of rational numbers Q, and the set of real algebraic numbers A to the set of real numbers R, so that

$$Z \subset Q \subset A \subset R$$

It can be shown that all these numbers can be represented in decimal form, those with a finite or repeating decimal expansion forming only a very small minority. It can also be shown that every decimal expansion, finite or infinite, represents some member of R. The set of all decimal expansions and the set of all real numbers are therefore equivalent.

The topological and other properties of the set R now form the starting point of most courses on *real analysis*. Intuitive ideas of limits and continuity are, it is true, to be found in most 'advanced level' school curricula, but the foundations of analysis lie firmly in the study of topology, and, whilst it is possible to develop considerable manipulative skill in the calculus without any formal topological background, it is the development of the formal concepts of topology that has provided a firm foundation by means of which much earlier work in analysis has been rigorously confirmed and upon which modern developments are rapidly being built. The value of an intuitive approach to mathematical concepts should not, however, be derided. Progress in mathematics is often the result initially of intuition; the rigorous proofs then follow. Intuition can sometimes lead to false conclusions, and it is for this reason that the value of an intuitive approach depends largely on the experience which underlies the intuition. In this book, the pattern of 'intuition first, formalization later' has been adopted, and it is hoped that the reader has obtained an intuitive grasp of some of the concepts which are properly the study of topology as well as having had an initial encounter with the kind of language and approach which formalization involves. Further study of topology will require an increasing familiarity with the formal concepts introduced in the later chapters of this book, and, for those who wish to pursue such studies, a selection (very incomplete) of books which may be read at this stage with profit is provided in the bibliography, page 182.

Historical Note

In the study of the development of man two main influences are recognized to have played and to continue to play decisive roles—*environment* and *heredity*. In a similar manner it is possible to discern both external and internal stresses at work shaping the genesis and growth of mathematical ideas. New branches of mathematics come into being, not because they are created overnight out of nothing by some individual genius, but because the soil has been prepared over the previous decades (or even centuries) and because some internal or external stress (or perhaps a combination of both) provides the appropriate impetus and motivation at the crucial point in time. More often than not, it is the case that several minds produce independently and almost simultaneously the germs of what subsequently develops into a new theatre of mathematical investigation. For this reason it is usually ill-advised to point to any one man as being the founder or inventor of any particular branch of mathematics.*

In the case of *topology* it is possible to see two particular and apparently quite separate areas of mathematics as providing much of the fertile soil out of which topology as a topic in its own right was to grow —the development of the *calculus* and the formulation of *non-Euclidean geometries*.

Some of the ideas which we today accept as commonplace in the study of the calculus can be traced back to the attempts of the Greeks to determine areas and volumes by the *method of exhaustion*. Newton and Leibniz put the calculus on its modern footing through their work on *differentiation* and *integration* showing the link between the two processes, and by providing appropriate notation by means of which problems involving rates of change could be formulated mathematically. However, the expositions of Newton and Leibniz created a number of difficulties which neither they nor their immediate followers were able to resolve. Many of these difficulties centred upon the fundamental concept of *limit*. The limit concept, essential to a proper understanding of the calculus, is inexorably tied to the concept of *nearness*, and in the late seventeenth and early eighteenth centuries the various inter-

---

*For a full discussion of stresses in the historical development of mathematics see R. L. Wilder: *Evolution of Mathematical Concepts*, Wiley 1968 (Paper-back edition, Transworld 1974).

pretations of the term 'limit' led to a number of conflicting principles which could be resolved only by some new approach allowing the concept to be restated in an unambiguous and rigorous manner.

The motivation for much of the basic investigation of the fundamentals of the calculus came from a number of physical problems of which the *problem of the vibrating spring* is a typical example. A purely internal motivation arose from the discovery of *pathological curves* which highlighted the distinction between *continuity* and *differentiability* and undermined the purely visual approach to the study of functions. Such investigations brought into prominence the need for a much more sophisticated method for the abstract formulation of mathematical problems which was only to become available with the advent of Cantor's *theory of sets*.

One of the major constraints under which mathematicians of these earlier days had to work was the over-riding veneration paid universally to the geometry of Euclid. Until the stranglehold of purely Euclidean concepts could be broken there was little chance that developments necessary for continuing the investigations into the foundations of the calculus could take place. It was the geometries of Lobachevsky and Riemann that provided exactly the release from Euclidean thraldom that was needed. In particular, it had been universally accepted for many centuries not only that Euclidean geometry was founded upon an unshakable axiomatic basis but also that it uniquely represented the real world in which all physical problems were assumed to arise. If the various non-Euclidean geometries had proved to be nothing more than impractical mathematical abstractions, it is doubtful if the necessary break from a purely Euclidean concept of space would have been achieved at the crucial moment; it was the realization that some of these geometries applied to practical and easily visualizable situations, such as geometry on the surface of a sphere, that set men's minds free and led to rapidly developing investigations into the nature of spaces, which continue until the present day.

The theory of point sets revolutionized the whole approach to the investigation of the nature of spaces by enabling it to be carried out in terms of sets of points having certain prescribed properties, and the rise of *functional analysis*, which led to the introduction of Hilbert and Banach spaces, underlined the importance of this mode of approach. The Euclidean distance function was, of course, used extensively in the definition of the important concept of *neighbourhood*, but early in this present century Hausdorff built up a theory of abstract spaces using a definition of neighbourhood presented entirely in set theoretic terms and not dependent upon the introduction of a metric.

The purely combinatorial aspects of topology may be said to go back to some of the geometrical work of Leibniz in which he sought to formulate basic geometrical properties of figures in terms of location rather than magnitude. Some seventy years later Euler was concerning himself with the relation between the numbers of edges, faces and vertices of closed convex polyhedra and also with the famous Koenigsberg Bridge problem. During the nineteenth century Moebius and Riemann pioneered the detailed study of surfaces and, in particular, Riemann linked the study of functions with the theorems of what was at the time known as *analysis situs*.

A further bonus arising from the liberation from earlier Euclidean restrictions was the extension of mathematical investigations to spaces of more than three dimensions. Much of the early work on combinatorial topology had been confined very largely to surfaces. A much more general attack on the combinatorial theory of geometrical figures was carried out at the close of the nineteenth theory by Poincaré. His study of configurations in higher-dimensional spaces was not, however, motivated entirely by theoretical considerations. He was especially interested in the qualitative theory of differential equations and this led him to investigate the structure of four-dimensional surfaces used in the representation of functions of complex variables, and hence to the systematic study of n-dimensional geometry.

Poincaré was not strictly concerned with the study of topological invariants. The concept of invariants under transformations in Euclidean space can be traced back to Desargues in the seventeenth century, whose work in this field was itself a development of earlier considerations of perspective in Renaissance art. The resulting geometry, known as *projective geometry*, together with the non-Euclidean geometries of Lobachevsky and Riemann were the subject of intensive investigation by Klein in the latter part of the nineteenth century. In particular, Klein concerned himself with the question of the consistency of non-Euclidean geometries and developed the idea that each geometry can be characterized by an appropriate group of transformations. This characterization was first presented on the occasion of his admission to the Faculty of Philosophy at the University of Erlangen in 1872 and is known as the *Erlangen Programme*.

All these various threads are drawn together in the study of topology. In each case the initial development has involved *generalization* which has in turn established links between different areas of mathematics; thus, the various generalizations have resulted in a considerable *unification* of mathematical concepts. In particular, conflicts between geometry and analysis have been resolved by a proper axiomatization of the concept of *space*. This is a not untypical way in which mathematics has developed over the centuries. Problems arise in a number of apparently distinct areas none of which can be adequately solved until a new methodology becomes available. Once such a methodology is developed the previously distinct problems are seen to have common foundations, threads are drawn together, and a new branch of mathematics is born. This is a continuing process, and can be found within topology itself. Once the new branch appears, then further processes of unification become possible as it interacts with other branches not previously involved in its genesis. Thus, there has been considerable fusion between topology and algebra, giving rise to a new system, *algebraic topology*, which has very substantially influenced the development and teaching of algebra in recent times. Indeed, it would be difficult to discover at the present time a branch of mathematics which could be said with any degree of confidence to be entirely independent of any topological considerations.

A Selection of
Exercises and Problems

These exercises and problems are designed primarily for the reader who, having completed a perusal of the whole book, would like to test his understanding of the material. They are not designed to test or to improve technique in any way. Some of them are intended to encourage thought or discussion, and certainly for these there is not necessarily any one ideal 'solution'. For this reason, 'answers' are not provided. Clearly, there are specific correct answers to many of the questions, but, rather than ask the reader to check his own work mechanically against printed solutions, it is hoped that he will satisfy himself as to the validity of his work by subsequent reference back to the text, by re-working the same questions independently after a lapse of time and comparing results, and (in the last resort) by seeking out some suitable person who can correct his work and discuss problems which arise out of the solutions actually obtained (whether right or wrong).

1 Demonstrate that translation is the only rigid transformation permitted in the orientated geometry described on page 9.

2 Show by a direct formula method that reflection of any two points about a given straight line in a plane preserves the distance (in the conventional sense) between the points.

3 Which of the following pairs are necessarily pairs of equivalent figures in the similarity geometry described on page 10.

   (a) two triangles of equal area,
   (b) two rhombi of equal area,
   (c) two rhombi of unequal area,
   (d) a rectangle and a square of equal area,
   (e) two rectangles of equal perimeter length,
   (f) two cones of equal volume,
   (g) two regular tetrahedra?

4 Which of the following statements are true:

   (a) each affine equivalence class is a subset of a similarity equivalence class,

172

(b) the set of all affine transformations is a subset of the set of all similarity transformations,

(c) each similarity equivalence class is a subset of a projective equivalence class,

(d) the set of all projective transformations is a subset of the set of all affine transformations,

(e) the set of all isometries is a subset of the set of all topological transformations?

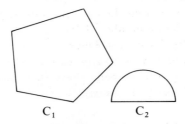

Figure A

5 Two closed 'contours' C_1 and C_2 are shown in Figure A drawn on a plane surface. Which of the following properties are geometric, which are topological, and which are neither geometric nor topological:

(a) the two contours define respectively a pentagon and a semi-circle,

(b) the area enclosed by C_1 is larger than that enclosed by C_2,

(c) no one point of the plane is inside both C_1 and C_2,

(d) C_1 and C_2 do not intersect,

(e) the diameter of the semi-circle defined by C_2 is equal in length to one of the sides of the pentagon defined by C_1,

(f) C_1 consists entirely of straight lines, whilst C_2 consists of a straight line and a curved line,

(g) C_1 and C_2 are on one and the same surface,

(h) C_1 and C_2 are drawn in black upon a white background,

(i) a line joining a point inside C_1 to a point inside C_2 will cross C_1 and C_2 at least once each.

(j) the areas enclosed by C_1 and C_2 together form only a part of the total surface upon which they are drawn,

(k) C_1 has more vertices than C_2,

(l) C_1 lies to the left of C_2?

6 What is the least number of continuous non-self-intersecting closed curves which may separate the surface of:

(a) a two-fold torus,

(b) a sphere,

(c) a sphere with a hole in its surface?

7 What is the genus of the surface of:

(a) a piece of wood with four screw holes right through it,

(b) a hockey stick,

(c) a ladder having exactly seven rungs,

(d) a frame of a tennis racquet?

8 Take several long narrow strips of paper and form a number of continuous bands by pasting the two ends of a strip to each other in each case after a differing number of 180° twists. Cut each band down its centre as partially depicted in Figure B, and investigate the effects of the various differing twists. In particular determine which of the original bands are one-sided and which cuts entirely separate the original surface. Discuss your results.

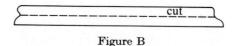

Figure B

9 Distinguish between the 'connectivity of a surface' and a surface 'being connected'.

10 To which of the following surfaces does the statement 'every continuous non-self-intersecting closed curve belongs to the null homotopy class' apply·

(a) a torus,

(b) a sphere,

(c) a Klein bottle?

11 What is the rank of:

(a) a kettle without its lid,

(b) a T-junction of piping,

(c) a ten-hole Meccano strip,

(d) a sphere with 3 handles and 4 holes?

In (a) to (c) assume that the item is made of 'infinitely thin' metal. Think of a variety of every-day objects and determine the respective rank of each.

12 Form a polyhedron by taking a point outside the centre of each face of a cube and joining each of these points to all the vertices of its corresponding face. Determine V, E, and F for this polyhedron. Repeat the

process, and again determine V, E and F. Verify that $V - E + F$ remains unchanged.

13 Which of the maps having values of V, E and F given below can be drawn upon the surface of a sphere:

    (a) $V = 8$, $E = 18$, $F = 10$,
    (b) $V = 5$, $E = 12$, $F = 3$,
    (c) $V = 6$, $E = 11$, $F = 7$,
    (d) $V = 12$, $E = 16$, $F = 5$,
    (e) $V = 23$, $E = 81$, $F = 60$?

Upon what surfaces can these maps be drawn if not on a sphere?

14 It is possible to brush a 'hairy' sphere so that it has only one singular point. Sketch s··ch a brushing, name the type of singular point, and state its index.

15 Figure C depicts the ground floor plan of a factory with gaps indicating doorways. Is it possible to make a tour of the factory starting

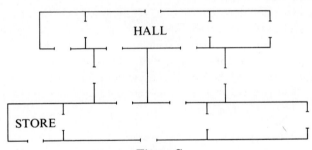

Figure C

from outside and passing through each doorway exactly once? Suppose that it was desired to make a similar tour starting inside the factory and without leaving the building. Indicate where doorways would have to be provided for this to be possible

    (a) starting in the hall and finishing in the store,
    (b) starting and finishing in the hall,
    (c) starting and finishing in any room whatsoever.

16 From the non-regular map of Figure D derive the regular map which has the least number of additional arcs and vertices.

17 The surface of a unit sphere in n-dimensional Euclidean space may be denoted by the expression

$$\sum_{i=1}^{n} x_i^2 = 1.$$

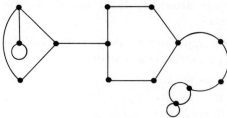

Figure D

Give a specific formula in the same form to define a curve homeomorphic to a Jordan curve.

18 The following is the major part of an alternative method of proving (by exhaustion) that if a line segment is divided into small segments by points arbitrarily labelled 0 or 1 (see page 86), then, if the number of 0's at the original end-points is odd (even), the number of complete segments is odd (even).

Consider a segment labelled 00. If this is divided into two by a point labelled 0, then no complete segment is formed; if by a point labelled 1, then two complete segments are formed. In either case the change in the total number of complete segments is even. Consider a segment labelled 01. If divided by a point labelled either 0 or 1, then there is still only one complete segment, again giving an even (zero) change . . . .

(a) Complete this proof by exhaustion.
(b) Why is it not necessary to consider segments originally labelled 10 and 11 as well as those already considered
(c) Try to complete the corresponding two-dimensional proof (using triangles) in a similar way. Why is this not possible ?

19 Sketch a two-fold torus and indicate how cuts may be made in order to obtain the corresponding plane diagram. By drawing an appropriate map upon the plane diagram, confirm the number of colours both necessary and sufficient for the colouring of maps on a two-fold torus.

20 Obtain the plane diagram of the interwoven surface consisting of the surfaces of two spheres, one inside the other, cut and then rejoined (as depicted in Figure *E*) so that the outer sheet of the 'Western' hemisphere joins the inner sheet of the 'Eastern' hemisphere and vice-versa along the lines *AB* and *CD*. Hence show that the interwoven surface is topologically equivalent to the surface of a torus.

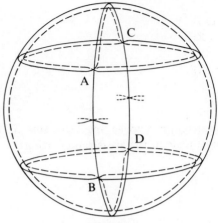

Figure E

21 Any surface homeomorphic to a sphere with p handles has $\chi = 2 - 2p$. Prove this by considering flow on a sphere with p sources and p sinks, and determining the index. Prove it also by considering only vortices in the flow.

22 Which of the following statements are true:

(a) a bicontinuous transformation is continuous,

(b) a transformation which is continuous at every point x in its domain is bicontinuous,

(c) a transformation, whose inverse is continuous, is continuous,

(d) a transformation, whose inverse is continuous, is continuous at some point in its domain,

(e) a transformation which is continuous at at least one point in its domain has an inverse which is continuous at at least one point in its domain?

23 If the universal set is taken to be the first twelve letters of the alphabet, and if $X = \{a, b, c, d\}$, $Y = \{d, e, f, g, h, i, j\}$, and $Z = \{i, j, k\}$, write down the members of:

(a) $\mathscr{P}(X)$,

(b) $X \cup Y$,

(c) $X \cap Y$,

(d) $X \cap Z$,

(e) Y',

(f) $X' \cap Y$,

(g) $Y \cap Z'$,

(h) $Z - Y$,

(i) $(X \cap Y) \cup Z$,

(j) $Y \cap (X \cup Z)$,

(k) $(X \cup Y)'$,

(l) $X \cap (Y \cap Z)'$,

(m) $X \times Z$,

(n) $X \times Y' \times Z$.

24 Draw a Venn diagram representing three sets X, Y, Z, no pair of which are disjoint. Shade in the areas representing:

(a) $X \cap Y$,

(b) $(X \cup Y)'$,

(c) $Y' \cup Z'$.

Hence, determine the simplest form of expressing

$$[(X \cap Y) \cup (X \cup Y)' \cup Y' \cup Z']'.$$

If $Y \subseteq Z$, what further simplification is possible?

25 If $X = \{a, b, c, d, e\}$ and $Y = \{1, 2, 3, 4, 5\}$, which of the following functions $f : X \to Y$ are injections and which are bijections:

(a) $f = [(a, 1), (b, 3), (c, 1), (d, 2), (e, 2)]$,

(b) $f = [(a, 1), (b, 3), (c, 2), (d, 5), (e, 4)]$,

(c) $f = [(a, 2), (b, 4), (c, 1), (d, 5), (e, 3)]$,

(d) $f = [(a, 1), (b, 4), (c, 3), (d, 2), (e, 5)]$,

(e) $f = [(a, 5), (b, 3), (c, 1), (d, 2), (e, 3)]$,

(f) $f = [(a, 5), (b, 5), (c, 5), (d, 1), (e, 2)]$,

(g) $f = [(a, 5), (b, 3), (c, 4), (d, 2), (e, 1)]$?

In the case of each injection, write down the subset of Y which should be taken as the codomain of the function so that f may be a surjection.

26 A triangle is defined by the three points in $\mathbf{R}^2$ represented by $(1, 1)$, $(0, 0)$, $(-1, 1)$. Find functions which will transform it into the triangle defined by $(3, -3)$, $(1, -5)$, $(3, -7)$.

27 If $f(z) = z^2 - z$, where $z = x + iy$, find the image of the rectangle defined by the points in the complex plane $(-1, 1)$, $(-1, 0)$, $(1, 0)$, $(1, 1)$.

28 If $f(z) = (z - i)/(2z + 1 + i)$, where $z = x + iy$, determine the image of the axis $(x, 0)$, and, in particular, of the set $\{(-\infty, 0), (0, 0), (\infty, 0)\}$.

29 If C is a circle passing through the point $(1, 0)$, enclosing the point $(-1, 0)$, and having its centre in the left-upper-half-plane, find the image of C when $f(z) = \frac{1}{2}(z+1/z)$. What difference can be observed in $f(C)$ when C encloses the whole of the unit circle and when it does not ?

30 Figures F and G depict two functions $f : \mathbf{R} \to \mathbf{R}$. In each case determine what subsets of $\mathbf{R}$ may be taken as domains or codomains of f in order that the inverse function f^{-1} may be defined.

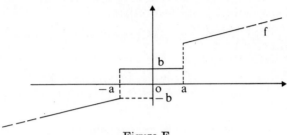

Figure F

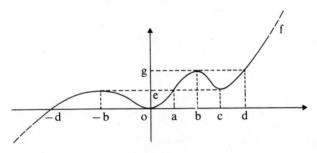

Figure G

31 If X is the subset of $\mathbf{R}^2$ in and on the unit circle centered at $(0, 0)$, and if f is the function which rotates all points of X about $(0, 0)$ through 90 degrees clockwise and all points of $\mathbf{R}^2 - X$ through 90 degrees counter-clockwise, for what subset of $\mathbf{R}^2$ is f continuous and for what subset is it discontinuous ? At points of discontinuity, for what values of ε are there no corresponding values of δ ?

32 Show that the set of all four-figure binary numbers with the metric d defined as the number of changes of digits required in going from one binary number to another is a metric space.

33 Let F be the set of all continuous real functions with domain $[0, a]$. Which of the following are metrics on F:

   (a) $d(f_1, f_2)$ is the maximum value of $\left|f_1(x) - f_2(x)\right|$ for $x \in [0, a]$,
   (b) $d(f_1, f_2) = \int_0^a \left|f_1(x)\right| - \left|f_2(x)\right|$,
   (c) $d(f_1, f_2) = \int_0^a \left|f_1(x) - f_2(x)\right|$,
   (d) $d(f_1, f_2) = \int_0^a \left|f_1(x) . f_2(x)\right|$?

34 If $X = \{a, b, c, d, e, f\}$, determine which of the following collections are topologies on X:

(a) $\{\{a\}, \{b\}, \{a, b\}, \{a, d, e\}, \{a, b, d, e\}, X\}$,

(b) $\{\varnothing, \{a\}, \{a, b, c\}, \{b, c, d\}, \{d, e, f\}, X\}$,

(c) $\{\varnothing, \{a\}, \{c\}, \{e\}, \{a, c\}, \{a, e\}, \{e, f\}, \{a, c, e\}, \{a, e, f\}, \{c, d, e\}, \{c, e, f\}, \{a, c, d, e\}, \{a, c, e, f\}, \{a, c, d, e, f\}, X\}$,

(d) $\{\varnothing, \{a\}, \{a, c\}, \{c, e\}, \{a, c, e\}\}$,

(e) $\{\varnothing, \{b\}, \{b, d\}, \{b, d, f\}, X\}$,

(f) $\{\varnothing, \{a\}, X\}$,

(g) $\{\varnothing, \{a\}, \{f\}, \{a, f\}, \{a, c, f\}, \{a, c, d, f\}, \{a, b, c, d, f\}, X\}$,

(h) $\{\varnothing, \{a\}, \{b\}, \{e\}, \{f\}, \{a, b\}, \{a, e\}, \{a, f\}, \{b, e\}, \{b, f\}, \{e, f\}, \{a, b, e\}, \{a, b, f\}, \{b, e, f\}, \{a, b, e, f\}, \{a, b, c, e, f\}, \{a, b, d, e, f\}, X\}$?

35 If $X = \{a, b, c, d, e\}$ and $\mathscr{T}$ is a topology on X comprising the collection

$$\{\varnothing, \{a\}, \{a, b\}, \{a, b, e\}, \{a, c, d\}, \{a, b, c, d\}, X\}$$

write down all the closed subsets of X. Find also:

(a) $\{a\}^-$,

(b) $\{b\}^-$,

(c) $\{c, e\}^-$,

(d) int $\{a, b, c\}$,

(e) ext $\{a, b, c\}$,

(f) bdy $\{a, b, c\}$.

36 Which of the following sets are connected:

(a) the circumference together with the interior of a circle, but with the centre point removed,

(b) $\{x : x \in \mathbf{R} \text{ and } 0 \leqq x < 2, 2 < x \leqq \infty\}$,

(c) a single point $x_0 \in \mathbf{R}^3$,

(d) the set of all points on all circles in $\mathbf{R}^2$ having centres at $(0, 0)$ and radii r, $r \in \mathbf{Q}$,

(e) the union of all connected subsets of a set, no two of which are separated,

(f) $f(X)$, if $f : X \to Y$ is continuous and X is connected,

(g) the null set $\varnothing$,

(h) $\{(x, y) : x, y \in \mathbf{R} \text{ and } x+y < 1\}$?

In each case where the set is not connected find a separation of the set.

37 Which of the following sets are compact:

(a) the interval $]a, b]$ of $\mathbf{R}$, $a \neq b$,

(b) a finite subset of a topological space $(X, \mathscr{T})$,

(c) $f(X)$, if $f: X \to Y$ is continuous and X is compact.

(d) the indiscrete topology on $\mathbf{R}^2$,

(e) the discrete topology on $\mathbf{R}^2$,

(f) a closed subset of a compact set ?

38 Which of the following sets are complete:

(a) the set $\mathbf{Z}$ of all integers,

(b) $\{x : x \in \mathbf{R} \text{ and } 0 < x \leq 1\}$,

(c) the set $\mathbf{Q}$ of all rational numbers,

(d) the set $\mathbf{R} - \mathbf{Q}$,

(e) the empty set $\varnothing$,

(f) $\{(x, y) : x, y \in \mathbf{R} \text{ and } 0 \leq x \leq 1, 1 \leq y \leq 2\}$?

Consider why the completeness of $\mathbf{R}$ implies the existence of irrational numbers.

39 Discuss briefly the concept of *continuity* on the basis of

(a) an intuitive approach,

(b) a metric space approach,

(c) a topological space approach.

40 Discuss the concept of *neighbourhood* and, in particular, consider how it provides an extremely useful conceptual link between the concept of a *metric space* and that of a *topological space*.

Bibliography

Blackett, D. W.   *Elementary Topology* (Academic Press).

Brown, R.   *Elements of Modern Topology* (McGraw Hill).

Cairns, S. S.   *Introductory Topology* (Ronald Press).

Hocking, J. G. and Young, G. S.   *Topology* (Addison–Wesley).

Kelley, J. L.   *General Topology* (Van Nostrand).

Lefschetz, S.   *Introduction to Topology* (Princeton University Press).

Lipschutz, S.   *Theory and Problems of General Topology* (Schaum).

McCarty, G.   *Topology* (McGraw-Hill).

Manheim, J. H.   *The Genesis of Point Set Topology* (Pergamon).

Mansfield, M.   *Introduction to Topology* (Van Nostrand).

Mendelson, B.   *Introduction to Topology* (Blackie).

Patterson, E. M.   *Topology* (Oliver and Boyd).

Pitts, C. G. C.   *Introduction to Metric Spaces* (Oliver and Boyd).

Simmons, G. G.   *Introduction to Topology and Modern Analysis* (McCraw-Hill).

Sze-Tsen Hu.   *Elements of General Topology* (Holden-Day).

Wallace, A. H.   *An Introduction to Algebraic Topology* (Pergamon).

Index